A CULTURAL HISTORY OF CHEMISTRY

VOLUME 3

A Cultural History of Chemistry
General Editors: Peter J.T. Morris and Alan J. Rocke

Volume 1
A Cultural History of Chemistry in Antiquity
Edited by Marco Beretta

Volume 2
A Cultural History of Chemistry in the Middle Ages
Edited by Charles Burnett and Sébastien Moureau

Volume 3
A Cultural History of Chemistry in the Early Modern Age
Edited by Bruce T. Moran

Volume 4
A Cultural History of Chemistry in the Eighteenth Century
Edited by Matthew Daniel Eddy and Ursula Klein

Volume 5
A Cultural History of Chemistry in the Nineteenth Century
Edited by Peter J. Ramberg

Volume 6
A Cultural History of Chemistry in the Modern Age
Edited by Peter J.T. Morris

A CULTURAL HISTORY OF CHEMISTRY

IN THE EARLY MODERN AGE

VOLUME 3

Edited by Bruce T. Moran

BLOOMSBURY ACADEMIC
LONDON • NEW YORK • OXFORD • NEW DELHI • SYDNEY

BLOOMSBURY ACADEMIC
Bloomsbury Publishing Plc
50 Bedford Square, London, WC1B 3DP, UK
1385 Broadway, New York, NY 10018, USA
29 Earlsfort Terrace, Dublin 2, Ireland

BLOOMSBURY, BLOOMSBURY ACADEMIC and the Diana logo are trademarks of
Bloomsbury Publishing Plc

First published in Great Britain 2021
Paperback edition published in 2025

Copyright © Bloomsbury Publishing Plc, 2025

Cover design: Rebecca Heselton
Cover image © British Library Board. All Rights Reserved/Bridgeman Images

All rights reserved. No part of this publication may be reproduced or transmitted
in any form or by any means, electronic or mechanical, including photocopying,
recording, or any information storage or retrieval system, without prior permission in
writing from the publishers.

Bloomsbury Publishing Plc does not have any control over, or responsibility for, any
third-party websites referred to or in this book. All internet addresses given in this
book were correct at the time of going to press. The author and publisher regret any
inconvenience caused if addresses have changed or sites have ceased to exist, but
can accept no responsibility for any such changes.

A catalogue record for this book is available from the British Library.

A catalog record for this book is available from the Library of Congress.

ISBN: PB: 978-1-3505-5208-1
        Pack: 978-1-3505-5229-6
        ePUB: 978-1-3502-5151-9
        ePDF: 978-1-3502-5150-2

Series: The Cultural Histories Series

Typeset by Integra Software Services Pvt. Ltd.
Printed and bound in Great Britain

To find out more about our authors and books visit www.bloomsbury.com
and sign up for our newsletters.

CONTENTS

LIST OF ILLUSTRATIONS   vii

SERIES PREFACE   xii

Introduction: Chemistry, Shifting Meaning, and Shapes of
Experience in the Early Modern Era   1
Bruce T. Moran

1   Theory and Concepts: Conceptual Foundations of Early
Modern Chymical Thought and Practice   23
Lawrence M. Principe

2   Practice and Experiment: Cultures of Chymical Analysis   41
Joel A. Klein

3   Laboratories and Technology: Chymical Practice and
Sensory Experience   67
Donna Bilak

4   Culture and Science: The Development and Spread of Chemical
"Knowledges" across Evolving Cultures and Communities   89
Andrew Sparling

5   Society and Environment: The Social Landscape of Early
Modern Chemistry   109
William Eamon

6 Trade and Industry: Chemical Economies and the Business
   of Distillation   141
   Tillmann Taape

7 Learning and Institutions: Chymical Cultures at Courts
   and Universities   171
   Margaret D. Garber

8 Art and Representation: Skepticism and Curiosity for the
   Alchemist at Work   199
   Elisabeth Berry Drago

NOTES   230
BIBLIOGRAPHY   232
LIST OF CONTRIBUTORS   255
INDEX   258

LIST OF ILLUSTRATIONS

0.1 Frontispiece from Giambattista della Porta, *Natural Magick* (1658). Credit: PD-US Wikimedia Commons 3

0.2 The library of Duke August. From Merian, *Topographia und eigentliche Beschreibung … Braunschweig und Lüneburg* (Frankfurt am Main, 1654/1658). Credit: Herzog August Bibliothek Digital Library 5

0.3 Frontispiece to Francis Bacon's *Instauratio magna* showing ships at sea sailing through the Pillars of Hercules to gain knowledge through exploration and by surpassing traditional learning. Credit: Wellcome Collection, London. CC BY 19

2.1 Title page of an early printed book with works attributed to Geber (Strasbourg, 1531). Credit: Science History Institute, Philadelphia 45

2.2 A man at an assaying furnace. From Georg Agricola, *De re metallica* (Basel, 1556). Courtesy Wellcome Collection, CC BY 4.0 46

2.3 Title page from Hieronymus Brunschwig's *Liber de arte distillandi* (Strasbourg, 1512). Othmer Library, Science History Institute, Philadelphia 49

2.4 Woman with bellows, from the title page of *Von allen geprenten Wassern* (Nurenburg, 1530). National Library of Medicine 50

2.5 Daniel Sennert. Line engraving by M. Merian (1628). Wellcome Collection, London CC BY 58

3.1 Interior of a Laboratory with an Alchemist (ca. 1640–1670) by David Teniers the Younger. Oil on canvas. Othmer Library, Science History Institute, Philadelphia ... 72

3.2 "The Alchemist" (after 1558) by Pieter Bruegel the Elder and Philip Galle. Engraving; first state of three. The Metropolitan Museum of Art, New York ... 74

3.3 "Common Furnace" by Nicaise LeFèvre, *Compendious Body of Chymistry* (1662). Othmer Library, Science History Institute, Philadelphia ... 76

3.4 "Wind Furnace" by Nicaise LeFèvre, *Compendious Body of Chymistry* (1662). Othmer Library, Science History Institute, Philadelphia ... 77

3.5 "Sublimating Furnace" by Nicaise LeFèvre, *Compendious Body of Chymistry* (1662). Othmer Library, Science History Institute, Philadelphia ... 78

3.6 "Athanor and Distillation Furnace" by Nicaise LeFèvre, *Compendious Body of Chymistry* (1662). Othmer Library, Science History Institute, Philadelphia ... 79

3.7 Emblem III by Michael Maier, *Atalanta fugiens* (1618). Othmer Library, Science History Institute, Philadelphia ... 85

4.1 Portrait of Paracelsus by the monogramist AH (ca. 1538). Reproduction 1927. Credit: Wellcome Collection, London CC BY ... 96

5.1 Pharmacy Transformed by Chymistry. Traditionally the apothecary practiced the simplest form of chemistry, grinding and mixing herbs and minerals to fill physicians' prescriptions. With the introduction of distillation and other chemical technologies the apothecary's art became more specialized and complex, as suggested by this depiction of a distilling oven with conical sheet metal condenser heads from Hieronymus Brunschwig's *Liber de arte distillandi de compositis* (1500). Wellcome Collection, London CC BY ... 113

5.2 Chymical technologies spread. Etching was first used to decorate armor and steel weapons, but by the sixteenth century was so common that it was also used to decorate everyday objects, such as butchers' and barrel-makers' knives. This cooper's knife was made in Germany in 1702. Victoria and Albert Museum, London ... 118

LIST OF ILLUSTRATIONS ix

5.3 Chymistry in the courts. Engraving from the *Nova reperta* (*New Inventions of Modern Times*) by Jan Collaert, after Jan van der Straet (Stradanus), depicting workers at a distillation laboratory in the court of Francesco I d'Medici in the Uffizi palace in Florence. The illustration depicts workmen busy at various alchemical tasks, and in the foreground is a scene of a distillation experiment overseen by court alchemist Sisto de Bonsisti, who peers at a book of alchemical secrets. Grand Duke Francesco looks over his shoulder. Jan van der Straet (1523–1605). Copper engraving. The Metropolitan Museum of Art, New York   127

5.4 The alchemist's image. This painting, titled *The Alchemist's Experiment Takes Fire*, by the Dutch painter Hendrik Heerschop, depicts a flask exploding while the alchemist is doing a transmutation experiment. The alchemist's tattered clothing (and that of his family in the background) is meant to show the futility of the alchemist's quest. Images depicting the fraudulent alchemist were common in the sixteenth century. Science History Institute, Philadelphia   129

5.5 A chymical revolution in the New World. The patio process of amalgamation invented by Bartolomé de Medina was put to use on a large scale in Mexico. This painting by Pietro Gualdi (1846) depicts the Hacienda Nueva de Fresnillo during silver reduction through the patio process. The dark-colored disks on the patio floor contain a semiliquid mass of pulverized, low-grade silver ore that has been mixed with mercury, ore masses that were central to the patio process, which was developed in sixteenth-century New Spain and used for approximately 350 years. Wikimedia Commons   136

6.1 Distilling furnace with an air-cooled alembic known as a *Rosenhut* ("rose hat"). Hieronymus Brunschwig, *Liber de arte distillandi de simplicibus* (1500), fol. 4r. Bayerische Staatsbibliothek   147

6.2 Title page illustration from Michael Schrick, *Von allen gebranten wassern* (Ulm, 1498). Bayerische Staatsbibliothek   148

6.3 Water-cooled still known as a "moor's head," suitable for distilling alcohol. Brunschwig, *Liber de arte distillandi de compositis* (1512), fol. 21r. Bayerische Staatsbibliothek   150

6.4 Illustrations and descriptions of distillation apparatus. Hieronymus Brunschwig, *Liber de arte distillandi de simplicibus* (1500), fol. 1v. Bayerische Staatsbibliothek   151

6.5	Elaborate water-cooling mechanisms. Walther Hermann Ryff, *Das New groß Distillier Buoch* (1545), fol. 26r. Bayerische Staatsbibliothek	157
6.6	Distillation of *aqua fortis*. Georg Agricola, *Vom Bergkwerck* (1557), p. 366. Bayerische Staatsbibliothek	159
6.7	Jan van der Straet, *Nova reperta* (ca. 1590), showing the new discoveries of the age, with distillation and guaiac wood in the bottom-right corner. Courtesy The Metropolitan Museum of Art	163
7.1	View of the city of Prague. Etching by Johannes Wechter (Aegidius Sadeler II, Antwerp/Prague, 1606). Harris Brisbane Dick Fund 1953, 53.601.10 (72–83), The Metropolitan Museum of Art, New York	179
7.2	Collaborative efforts in the Hapsburg orbit: title page of Oswald Croll, *Basilica chymica* (1609), illustrated by the court artist Aegidius Sadeler, edited and used by Johannes Hartmann for the latter's *laboratorium* course. Wellcome Collection, London CC BY	181
7.3	Boerhaave giving a lecture at the University of Leiden. Title page, Hermann Boerhaave, *De comparando certo in physicis* (Leiden, 1715). Wellcome Collection, London CC BY	197
8.1	Pseudo-Geber, illustration of water bath apparatus from the *Summa perfectionis magisterii*. Venice, 1542. Othmer Library, Science History Institute, Philadelphia	210
8.2	Jan and Caspar Luyken after Christoph Weigel, "Der Alchymist oder Goldemacher," ca. 1698. Engraving. Wellcome Collection, London CC BY	212
8.3	Adriaen van de Venne. *Rijcke-Armoede*, ca. 1630–1632. Oil on panel. Science History Institute, Philadelphia	214
8.4	Richard Brakenburgh, *An Alchemist's Workshop with Children Playing*, ca. 1670–1680. Oil on canvas. Science History Institute, Philadelphia	215
8.5	Hendrick Heerschop, *An Alchemist with His Assistant*, ca. 1660–1680. Oil on canvas. Science History Institute, Philadelphia	217
8.6	Mattheus van Helmont, *An Alchemist at Work*, ca. 1650–1680. Oil on canvas. Science History Institute, Philadelphia	218
8.7	Mattheus van Helmont, *The Alchemist*, ca. 1650–1680. Oil on canvas. Science History Institute, Philadelphia	218

LIST OF ILLUSTRATIONS

8.8 David Teniers the Younger, *Alchemist with Book and Crucible*, ca. 1640–1670. Oil on panel. Science History Institute, Philadelphia   219

8.9 *Le Plaisir des Fous*, Pierre Francois Basan after David Teniers the Younger. Science History Institute, Philadelphia   221

8.10 Cornelis Bega, *The Alchemist*, ca. 1663. Oil on panel. Getty Museum, Los Angeles/digital image courtesy of the Getty's Open Content Program   222

8.11 Cornelis Bega, *The Alchemist*, ca. 1663. Oil on canvas, mounted on panel. From the Collection of Ethel and Martin Wunsch/National Gallery, Washington, DC   223

8.12 Thomas Wijck, *Interior with an Alchemist*, ca. 1660–1677. Oil on panel. Wellcome Collection, London CC BY   225

8.13 Thomas Wijck, *Alchemist and Family*, ca. 1660–1670. Oil on canvas. Science History Institute, Philadelphia   226

8.14 Heinrich Khunrath, Plate 3 from *Amphitheatrum sapientiae aeternae* (*Amphitheater of Eternal Wisdom*). Hanau, Germany: Wilhelm Antonius, 1609. Othmer Library, Science History Institute, Philadelphia   228

SERIES PREFACE

A Cultural History of Chemistry examines the history of chemistry and its wider contexts from antiquity to the present. The series consists of six chronologically defined volumes, each volume comprising nine essays; these fifty-four contributions were written and/or edited by a total of fifty scholars, of ten different nationalities. Of Bloomsbury's many six-volume *Cultural Histories* currently in print, this is the first in the physical or natural sciences; it is also the first multivolume history of chemistry to appear since James Riddick Partington's four-volume *History of Chemistry* concluded more than fifty years ago. It is distinguished, among other qualities, by its endeavor to take the subject from antiquity right to the present day.

This is not a conventional history of chemistry, but a first attempt at creating a cultural history of the science. All cultures, including the various branches of natural science, consist of mixed constructs of social, intellectual, and material elements; however, the cultural-historical study of chemistry is still in an early stage of development. We hope that the accounts presented in these volumes will prove useful for students and scholars interested in the subject, and a starting point for those who are striving to create a more fully developed cultural history of chemistry.

Each volume has the same structure: starting with an interpretive overview by the volume editor(s), the eight succeeding chapters explore for each respective era in chemistry its theory and concepts; practice and experiment; laboratories and technology; culture and science; society and environment; trade and industry; learning and institutions; and art and representation. Readers therefore have the option to read multiple chapters in a single volume, thus learning about the cultures of chemistry in a single era; or they may prefer instead to read corresponding chapters across multiple volumes, learning about (e.g.) the art and representations of chemistry through the ages. Though the scope is global, major emphasis is placed on the Western tradition of science and its contexts.

Whether read synchronically or diachronically, in any multiauthor undertaking like this one readers will inevitably notice overlaps and repetitions, conflicting historical interpretations, and (despite the magnitude of the project) occasional gaps in coverage. These are inescapable consequences, but they actually offer advantages to the reader, both in making each chapter closer to self-contained and in demonstrating the dynamism of the discipline; like science itself, the study of its history is ever contested and incomplete.

Chemistry has been called the "central science," due to its fundamental importance to all the other physical and natural sciences. It is the archetypical science of materials and material productivity, and as such it has always been deeply embedded in human industry, society, arts, and culture, as these volumes richly attest. The editors and authors hope that *A Cultural History of Chemistry* will be of great interest and enjoyment not just to chemists and specialist historians of science, but also to social, economic, intellectual, and cultural historians, as well as to other interested readers.

Peter J.T. Morris and Alan J. Rocke
London (UK) and Cleveland (USA)

Introduction

Chemistry, Shifting Meaning, and Shapes of Experience in the Early Modern Era

BRUCE T. MORAN

Cultural historians know that context is key in uncovering the web of relationships underlying the meaning of thoughts and actions. In that respect, the relationships embedded within the history of early modern chemistry develop within a framework of distinct communities – elite as well as popular, Latinate as well as vernacular, textual as well as material – that preserve the norms and procedures of specific publics and traditions (Cooter and Pumfrey 1994). Exploring the cultural history of chemistry in the sixteenth and seventeenth centuries, then, requires paying attention, on the one hand, to the language, symbolic structures, and authorities that establish chemistry's identity and sense of place, and, on the other, to the networks, materials, and tools that fashion the general economic and social experience within which that identity is constructed. This is by no means an easy task, especially since the very definition of "chemistry" is, during this period, contested territory. Even catchall definitions inevitably run into boundary disputes. Regarding chemistry in terms of techniques that bring changes to nature seems safe; but in the early modern era something else does that as well, namely magic. What was called natural magic, to distinguish it from the demonic sort, comprised what we would today call natural knowledge. At the beginning of the sixteenth century the occult philosopher and magician Cornelius Agrippa defined its focus as the knowledge of "whole nature," being concerned with "the differing and agreement of things." In the hands of the *magus*, it produced "wonderful effects" by uniting the virtues of nature's parts (Agrippa 1898: book I, chap. 2). Near the century's close that definition, broadened by observational and experimental strategies of discovery, still remained normative. In 1589 the Neapolitan natural philosopher Giambattista della Porta also defined magic as an art that opened the "properties and qualities of hidden things." The magician, he added, was an artificer with a knowledge of nature, including metals, minerals, gems, and stones. He was cunning in the art of distillation and sublimation, and discovered thereby "many things profitable for the use of man." These he learned not in a "rude and homely manner," but with an acquaintance of "causes and reasons" (Porta 1957: 2–3) (Figure 0.1). Part of magic, in other words, was chemistry, and part of chemistry was magic. Writing the cultural history of chemistry in the early modern period is not a matter of filling in historical blanks with familiar terms and assumed discoveries. The challenge is to uncover the language, interpretations, and experiences by which the period made sense of itself.

SETTING THE STAGE: THE BIG PICTURE

The chronological focus of this volume from a European perspective is a period of unprecedented religious, intellectual, and socioeconomic change, one that challenges inherited categories and authorities and witnesses the emerging relevance of new social groups, disciplines, institutions, and methods of inquiry. This was an era of transition from the classical values of the Renaissance to attitudes and styles of learning with a more modern ring. The history of what might be called chemistry during this period is a history of various sorts of activities sometimes not specifically defined as chemical at all, but each embedded in particular economic and social realities, political events, philosophical interactions, and private affairs.

A snapshot of the life span of the alchemist, physician, lay theologian, and natural philosopher Theophrastus von Hohenheim, called Paracelsus, certainly one of the most significant figures in terms of both chemistry and medicine in the period's early years, gives us a good picture of some of the more prominent changes underway. When he was born, near the end of 1493, the Italian mathematician Lucca Pacioli was standardizing a method of accounting, later known as double-entry bookkeeping, which helped fuel a new market economy. Pacioli's friend Leonardo da Vinci was designing a colossal equestrian statue for the Sforza duke of Milan and imagining machines useful in war and manufactures. Further south, Michelangelo had just bought a large block of marble intending to carve an imposing statue of Hercules and was, along with Leonardo and many others, representing the world in three dimensions by employing what has been described as either an invention or a discovery, namely

FIGURE 0.1 Frontispiece from Giambattista della Porta, *Natural Magick* (1658). Credit: PD-US Wikimedia Commons.

perspective drawing. In 1493 Columbus had just recently arrived back in Spain from his first voyage and was composing a letter concerning his observations of the island he called *Hispana* (the island shared by the Dominican Republic and Haiti today), noting how easy it would be to conquer the people living there. To the north, Martin Luther was still a boy studying Latin, but, having learned to read and write, he would, during Paracelsus' lifetime, occasion many of the reforms in Christian ritual and belief that helped set in motion the Protestant Reformation. In pursuing those reforms, Luther, like many others, availed himself of what some have seen as the era's most important innovation, a change in the organization and communication of knowledge that shifted the world of learning from a culture of scribes to a culture of print.

The technology of printing with moveable type altered access to learning, ordered and organized what was read, stockpiled information, and expanded the opportunities for criticism among an increasingly literate public. Books changed habits of education, stimulated expression, and, for those willing to invest in their production, created new opportunities for making money. When Paracelsus came into the world the number of presses in Western Europe (many of the most important in Italy, Germany, and France) had grown from a single press in the German city of Mainz to roughly a thousand, and these already had churned out, in the space of about fifty years, books estimated in the millions (Eisenstein 1979: 43ff). Reference works of enormous size compiled excerpts gathered from already existing sources, managing information by means of classifications, cross-references, and diagrams (Blair 2010). With many more books, libraries became bigger and better organized. Spain, Italy, England, and France boasted major royal and ecclesiastical collections. In Central Europe, princely libraries stood out. One prince, Duke August the Younger of Braunschweig and Lüneburg, personally supervised his library's construction in the German town of Wolfenbüttel, his family's residence, presiding over the purchase and organization of what one observer estimated to be around 116,000 titles in 28,000 volumes (Figure 0.2). The library still offers readers today one of the most all-encompassing collections pertaining to the study of nature, including subjects related to chemistry, in the early modern period.

Some books excited attention by publishing reports of curiosities linked to recently discovered places and peoples. Many used pictures, especially woodblocks and copperplate engravings, to visualize what readers often had never seen before. Still others conveyed ideas that broke with long-accepted opinion. In 1541, the year that Paracelsus died, one of those books described a revolutionary astronomical system that placed the earth in motion around a stationary sun, summarizing ideas that were being worked out by the Polish–German mathematician and church administrator, Nicolaus Copernicus.

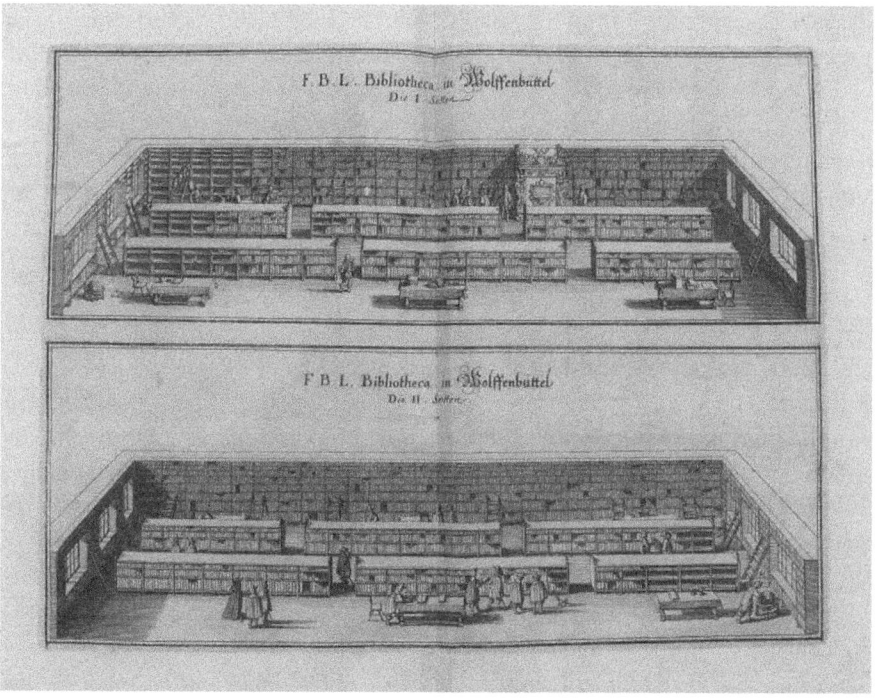

FIGURE 0.2 The library of Duke August. From Merian, *Topographia und eigentliche Beschreibung ... Braunschweig und Lüneburg* (Frankfurt am Main, 1654/1658). Credit: Herzog August Bibliothek Digital Library.

Copernicus' own text, *De revolutionibus orbium coelestium* (*Concerning the Revolutions of the Celestial Orbs*), appeared in print two years later, the same year in which there appeared one of the most important books published in the history of anatomy, *De humani corporis fabrica* (*Concerning the Fabric of the Human Body*), written by a twenty-eight-year-old Brussels-born physician teaching at the University of Padua, Andreas Vesalius.

The commercialization and the commodification of European society, with their attendant institutional and political changes, form the backdrop against which the production of knowledge by way of material creativity and new thinking concerning the operations of nature took place. The influx of gold and silver from the Americas fueled economic expansion and allowed some Europeans to invest in new trades and manufactures. Experience, experimentation, and quantification shaped inquiry and brought together technology, erudition, and artisanal know-how in the service of political ambition and the pursuit of profit. Competition for empire among major European states linked patronage with exploration. Europeans loved discoveries, especially when they produced marketable commodities. In Spain imperial mathematicians devised instruments and organized programs of astronomical observation to aid navigation. In addition, Spanish political and economic interests encouraged colonial expansion, motivated novel methods of gathering natural knowledge, and generated imaginative ways for verifying reports of uncommon and potentially useful things (Barrera-Osorio 2006; Bleichmar et al. 2009; Portuondo 2009).

New World plants and animals altered the landscape, as well as the cuisine, of Europe, but it was the gold and silver weighing down the hulls of returning ships that, once minted into coin, stimulated economic as well as social and political change. On the one hand, the new wealth advanced an evolving capitalist system of markets, private property, and exchange, and, on the other, enhanced the consolidation of political power among territorial princes and monarchs. In England it encouraged a surge of manufactures, paper and gunpowder mills, alum and copper factories, saltpeter works, and sugar refineries. The Dutch were particularly resourceful, creating joint-stock enterprises whose merchant fleets carried goods across the globe – spices, tea, porcelain, and cotton from Asia; sugar, tobacco, and furs from the Americas; and lumber, grain, cattle, and copper from Northern Europe. The most important venture was the Dutch East India Company, founded in 1602 with an initial capital outlay, gathered together from individual investors, of six and a half million florins. Dealing in quantity, these companies undersold their competitors, and, wherever possible, increased profits by altering and improving raw goods in mills, dye houses, refineries, and distilleries for sale to buyers from all over Europe. As Europe became wealthier, populations expanded. Estimates vary considerably, but conservative accounts have Rome at around 50,000 established residents in 1526 and 100,000 by the end of the century. Antwerp grew by a similar amount by 1560, and Amsterdam expanded from around 30,000 in the mid-1500s to near 200,000 a century later.

Investment and economic growth, however, did not equate to improved standards of living for everyone. Europe had not created the source of its own wealth, and the gold and silver of the new world prompted an era of extraordinary inflation. Old values began to shift as economic and political power gravitated toward merchant enterprise. In this context, alchemy and its products provided a means of commercial advantage and economic well-being. The German court physician, alchemist, and at one time commercial advisor to the Holy Roman Emperor Johann Joachim Becher encouraged his prince to turn away from a dependence upon landed wealth and create projects aimed at the creation of manufactured goods as a way of securing political authority. Alchemical processes and techniques, he argued, could thus help the prince engage the new money and market economy by being one of the driving forces of material production (Smith 1994).

In the history of chemistry Becher is usually remembered for his thinking concerning metals and for his theory of "phlogiston" (the subtle material stuff that he and others thought to be responsible for combustion and calcination). Yet, to the world that he knew and that knew him best, Becher was better known as a dedicated alchemist who believed steadfastly in the transmutation of metals and who pronounced alchemy's great value in advancing other arts (e.g. glass-making, dyeing, ink-making, the preparation of medicines, and the production of gunpowder). Alchemy was, he thought, the philosophic key to understanding the nature of metals and the processes of metallurgy. However, it also played a larger role within the emerging market economy. It was the means by which the merchant manufacturer produced the commodities that allowed money to change hands and that served both the power of the prince and the well-being of the civic community. By increasing manufactures it also increased the labor market, and labor, Becher thought, was the only true currency. This was a revolutionary role for alchemy – the means of creating genuine rather than artificial human wealth, the kind of wealth that produced equality through labor rather than giving rise to the disparities of proprietorship. While acknowledging that the production and consumption of "unnecessary luxuries" was one source of inequality, money itself, Becher reasoned, especially the sort based on the possession of gold and silver, was another. This sort of wealth, he thought, produced a masquerade that deceived the entire world. Much later Karl Marx described money as the supreme chemical binding agent of society, capable of transforming all individual inadequacies into their opposite, making the dishonest, good; the stupid, clever (Marx 1988). Although his remarks were more moral than social, Becher also observed the transformative power of money, reflecting that money could make fools into priests and scholars out of fools. It made the appearance of the honorable from what was in fact dishonorable, and made an intelligent man the servant of an imbecile. Labor, he mused, was the only "coin given to all men" (Becher 1669: 151), and while rank and privilege remained matters of birth, the combination of labor, manufacture, and practical alchemy (i.e. chemistry) helped to adjust society's economic imbalance.

Like many others of the time, Becher combined practical and experimental pursuits in the study of nature with a wide variety of other interests. In his case, these included medicine, moral philosophy, and political economics. Such a mixture of experience, sentiment, and belief are staples of cultural life, and sometimes join ways of thinking that, from a historical distance, seem conflicting or paradoxical. Then as now natural knowledge cohabited with everyday fears, superstition, and the raw desire for riches and power. To avoid cartoonish caricatures, making knowledge in the early modern world must never be separated from ugly realities. The methodological and experimental insights of Francis Bacon and René Descartes and the order of nature depicted by Newtonian physics shared historical space with the persecution of witches, the experience of warfare on a massive scale, and the shipping of over a million African slaves to the Americas.

Amalgams of motive and belief fashioned routine experience, but they also turned the wheels of discovery and the production of natural knowledge. What we might like to call early modern "science" was itself an amorphous set of judgments and practices shaped by devotion, conjecture, and authority on the one hand and reason and technical know-how on the other. It was not only reasoned mathematics and observation that produced Newton's theory of matter and his insights regarding terrestrial and celestial dynamics, but also his reading of ancient texts and a commitment to the omnipresence of God in the natural world. Many of the books that influenced Newton's alchemical thinking were published in the sixteenth and seventeenth centuries, and the ideas in many of those texts helped Newton link alchemy, experiment, and physics in the pursuit of a "sophic mercury" that would produce transmutations (Newman 2018). In the writings of Robert Boyle, empirical analysis and a corpuscularian-mechanical view of nature lived side by side with an intense personal spirituality that made the collection of experiences of contact with the supernatural and a fascination with reports of "second sight" (those who claimed to see into the future) altogether reasonable. Religious and evangelical sentiments drove also Boyle's involvement with the English East India Company in which he hoped to make use of the company's outposts to spread Christianity to indigenous peoples; and they also help explain his caution in regard to alchemy. In two instances of successful transmutations of lead into gold that he claimed to have directly witnessed, he detected the presence of the supernatural, and worried elsewhere that alchemical insights, including transmutation, might originate from malevolent spirits (Principe 1998; Hunter 2009).

WHAT IS CHEMISTRY AND WHERE DOES IT BELONG?

In tracing the characteristics of early modern chemistry we should bear in mind that modern-day distinctions regarding what counts as theory and what counts as practice, what counts as "science" and what does not, do not work well in the premodern world. In this sense particularly, cultural history allows the period to speak for and to define itself, taking note of its own use of terms, its own means of establishing and evaluating facts, and its own notions about what counts in the production and presentation of natural knowledge and technical acumen. Allowing the era to express itself, however, gives rise to some bothersome problems of terminology and classification. What we may be tempted to describe as the instruments, spaces, and practices of early modern *chemistry* often turn out to be represented by historical actors themselves as the tools, sites, and procedures of traditional *alchemy*. To make matters worse, historical actors differed in their characterization of terms like *alchemia*, *chemia*, and *chymia*, and attempts by historians to get around the complexity produce interpretations that more often ignore rather than explain disputes and native reasoning. For heuristic reasons, a good suggestion has been to use the term "chymistry" as a linguistic umbrella intended to preserve the old in the new and maintain an alchemical presence in the experimental philosophy of the era's "new learning" (Newman and Principe 1998). Yet the experience of what gets lumped together as chymistry was never one experience, and never reflected a unified structure of meaning. Neither were categories of application nice and tidy, with *chimia*, *chemia*, *chymia*, *chemica*, *chimica*, and *chymica* reserved for making medicines and *alchymia*, *alchimia*, *alchemia*, etc. reserved for fabricating metals and transmutation (Abbri 2000).

Discussing early modern chemistry, or chymistry, then, much depends upon regional usage and the associations fashioned in specific environments and circumstances. Even words that look similar, like *chymia* and *chemia*, often had different meanings dependent upon context. Today, people who make and study pottery refer to terms like "maiolica" and "majolica." These words too look almost identical, and, while in general discussions they are often used interchangeably, some ceramicists understand subtle differences between them relative to time and place. In the mid-seventeenth century, the German physician and polymath Hermann Conring expressed his own sensitivity to differences in the use of terms. He refused to refer to *chymia*, as some had done, to connect chemical medicine to ancient Greek practices and especially to the ancient sage Hermes Trismegistus. No tradition of chemical medicine could be attached to Hermes, he argued. Moreover, he continued, the Greeks only knew *chemia*, which they solely applied to the improvement of metals (Conring 1648: 15).

It may seem trivial, but the issue of names is important because it relates to disputes over how the emerging subject of chemistry ought to be defined and what it should contain. In ordering the world by means of categories, education in the early modern era focused repeatedly on distinctions between *genera* (classes of things that differed in kind) and *species* (those things that differed in number, or individuals, but not in kind). The *genus* "animals," for instance, is made up of *species* of dogs, tigers, and pigeons. But both animals and tigers might be *species* in relation to another, bigger, *genus*, say, "living things." For purposes of reasoning about any subject, these were crucial distinctions, and every student had drilled into his or her head that whatever was true for a *genus* was also true for its *species*; however – and this is the most important thing – it was logically wrong to claim the reverse. In arguing about what counted as "chemistry" in the early modern era, defining what sorts of experience, procedures, and explanations were species of the subject and what were parts of something else really mattered.

Early on, Paracelsus and his followers claimed terms like *chemia* or *chymia* as their own and made both part of the *genus* of medicine. Their view also implied a preternatural understanding of nature based in large part upon a vitalist cosmology shared, among other traditions, with Renaissance hermetism. For some, letting this view define the chemical art was to get the subject all wrong. One who especially felt this way was a German physician and schoolteacher named Andreas Libavius. He preferred the term "alchemy" to describe procedures for purifying, combining, and separating substances, for analyzing metals and minerals, and for making useful things from substances received from nature. Yet alchemy too amounted, in terms of definition, to a moving target, and not everyone agreed about what it actually was. Some assigned the art of alchemy to the preparation and transmutation of metals. Some thought of it solely in terms of making medicines. Some limited it further to making an elixir. Some thought of it as the knowledge of preparing salts. Some described it in purely mystical terms, following Hermes. Some, like Hieronymus Rubeus (i.e. Girolamo Rossi), wanted to separate the art of distillation from alchemy; others, like pseudo-Geber and Avicenna, thought that it produced waters and oils by means of distillation, and sublimated a mercurial elixir for purposes of transmutation. Some thought that Paracelsus was its true inventor. Some thought that the art of the alchemist was simply craft or cunning, and no art at all.

When the Italian mathematician, mechanician, and physician Hieronymus Cardanus (i.e. Girolamo Cardano) sought to describe what he called the *ars chymistica* (the art of the alchemist or chymist) he avoided the thorny business of theoretical definition altogether and preferred to call attention to the products of those artisans who knew how to bring changes to nature and make things. According to Cardanus, these artisans knew how to stretch glass into long strands and how to interweave glass with white threads. They also knew how to make false gems and artificial amber and how to engrave or etch images onto glass. They mixed, altered, and refined metals. They made white gold, or electrum (a gold–silver alloy). They purified camphor and produced waters and oils by means of distillations and extractions with alcohol, and they knew how to construct the instruments of distillation. In particular, they could create dissolving waters able to penetrate the smallest spaces between the material parts of bodies and thus could separate even the most intractable substances. Some of the things made by the practitioners of the *ars chymistica*, Cardanus admitted, where worthless or dubious, but other things were admirable, beautiful, health-giving, and efficacious. Some things, he added, were divine (Cardanus 1551: 269v–71r). Mundane or heavenly, to Cardanus, the *ars chymistica* was the skill or technique needed to make certain kinds of things from the stuff of nature.

To Libavius and others, however, simply making things did not comprise an art. Whatever was to be described as a "chemical art" needed a proper philosophical foundation, and ironically, it was in large part due to the efforts of Libavius to provide a foundation based in Aristotelian logic that the subject of *chemia* got a new life. In wresting away *chemia* from the grasp of Paracelsians and fakes, he declared that true *chemia* was *physica* (i.e. natural philosophy), and its theory and practice was the best way – better than any other art – to inquire about the operations of nature. As an art unburdened by secrecy and open to everybody, *chemia*, he argued, might replace alchemy as a subject capable of disclosing the powers residing within the objects of nature and, through public exchanges of knowledge, profit human society. Its language would be transparent and communicable. It could be taught and learned in schools, and its practitioner, the true chemist, would possess a public purpose anchored in social virtue. Significantly, Libavius added, the subject of *chemia* would then be an art that stood on its own disciplinary feet, and this meant that, in terms of traditional institutional norms, its definition would be independent of medicine (Moran 2007; Moran 2014).

CHEMISTRY, MEDICINE, AND *CHYMIATRIA*

This was a revolutionary suggestion. Nevertheless, it took a long time for the subject of "chemistry" to establish its true independence from the medical arts, and an even longer time for it to be considered relevant in establishing physical theory. Within the educational structure of the European university, most continued to think that the various theories and practices related to *chemia* or *alchemia* were part of the subject of medicine. In the sixteenth century the physician and naturalist Pier Andrea Mattioli wrote:

> I dare to say that no physician can be complete … who is not experienced in the most noble art of distilling: that not only elsewhere but also especially in chronic illnesses is to pay attention to those who are able to bind fast the power within metals. Neither are there lacking examples of great physicians who have brought forth great effects in curing illness by means of chemical remedies.
>
> (Mattioli 1561: lib. 4, 317)

In preparing chemical remedies Paracelsian physicians played a prominent role, and many began to shed the coat of mysticism associated with Paracelsus' original views. As they saw it, their task was to extract, by means of fire and especially through distillation, sublimation, and calcination, the subtle virtues of a particular substance that corresponded to the diseased organ in the body. Only then could an effective medicament be administered. However, those techniques were by no means new and it would be a mistake to think that the preparations of chemical medicines were exclusively the province of Paracelsian practitioners. Critics, in fact, were often of two minds. The imperial physician Crato von Krafftheim wrote that he did not disavow the medicines handed down straightforwardly and to good purpose by Paracelsus, but he did condemn "those malicious men and lying Paracelsian tricksters" who had made a show of the metamorphosis of gold and swindled others as a result. These, he declared, would simply never know "the divine art of medicine" (Krafftheim 1582: epistola). Others grounded the preparation of chemical medicines in ancient medical traditions, casting Paracelsian practices more generally into the cauldron of deceit. The Brandenburg court physician and professor of medicine at Frankfurt an der Oder Caspar Hofmann held that "the doctrine, method, and comprehension" of medicine supported by "reason, extended by observations, and confirmed by experience" was due to Hippocrates, and that the art had been brought to its most perfect state by "the methods of definition, division, and analyses" constructed by Galen. Nevertheless, he also observed that the remedies afforded by *chemia* and "the violence of the fire," if the true method of healing was observed, could be bestowed with honor and serve as a maidservant to medicine. However, there were, he warned, imposters, the "smokey sons of Vulcan," who had made a shipwreck of the practice of healing and had introduced monstrous errors and grotesque follies into medicine. Puffed up in their own ignorance, they made judgments in the name of Paracelsus and sucked out money from the purses of those eager for novelty (Hofmann 1726: 258–60, 264–8).

Paracelsian physicians in particular relied upon court rather than university appointments to establish social credibility for their medical theories and procedures. The French Paracelsian physician Joseph Du Chesne (Quercetanus) assembled a list of princes who sponsored chemical medicine. These included the Holy Roman Emperor and the king of Poland, as well as the archbishop of Cologne, the Duke of Saxony, the Landgrave of Hesse, the Margrave of Brandenburg, the dukes of Braunschweig and Bavaria, and the princes of Anhalt. The list missed the Duke of Neuburg, who later became Elector of the Palatinate, and Ott-Heinrich, whose physician, Adam von Bodenstein, was one of the first publishers of Paracelsus' works. With the financial support of the Calvinist prince Christian I of Anhalt-Bernburg, the court *medicus ordinarius* (chief medical representative) Oswald Croll constructed one of the most important expositions of Paracelsian remedies, the *Basilica chymica* (1609), and he dedicated the book to his prince. In Denmark, the professor of medicine and royal physician Petrus Severinus (1542–1602) brought Paracelsian ideas into line with academically acceptable philosophical traditions (Shackelford 2004), while at Paris, Paracelsian physicians at the court of Henry IV adopted chemical practices and engaged in heated argument with the Parisian university faculty of medicine concerning chemical medicines and their bearing upon ancient medical theory (Kahn 2007a; 2007b).

One term often used by those who practiced making chemical remedies in the early modern era was *chymiatria*, and in 1624 a city and court physician named Thomas Reinesius, who had earlier been a teacher at a gymnasium in the German city of Gera, offered an encompassing definition of what the term meant. Reinesius noted that *chymiatria*

> is that hidden knowledge of the philosophers of healing well by dissolving mixed natural bodies, coagulating the dissolved, and separating, by means of fire, the pure from the impure and [thereby] manifesting what is hidden in mixed bodies for the purpose of producing more agreeable, protective, and efficacious medicines. As such it holds and preserves an honest place as a worthy art, spreading its knowledge eternally to posterity in academies, gymnasia, pharmaceutical workshops, and in the colleges of all learned physicians.

By the time Reinesius commented on the subject of *chymiatria*, authors who had, in their own earlier writings, been at each other's throats could be lumped together in a pile of influential predecessors that Reinesius considered worthy of memory and praise either because they had "recognized, cultivated, and bequeathed ... certain pieces and fragments of the chemical art or had taught the art generally by means of precepts." His list was a veritable "Who's Who" of early modern notables in chemical medicine and included Conrad Gesner, Gerhard Dorn, Julius Caesar Scaliger, Jean Fernel, Johannes Albertus Wimpinaeus, Pietro Andrea Mattioli, Johannes Crato von Krafftheim, Felix Plater, Petrus Severinus, Thomas Muffet, Theodore and his son Jacob Zwinger, Joseph Du Chesne (Quercetanus), Martin Ruland, Raymund Minderer, Hieronymus Reusner, Giambattista della Porta, Johannes Hartmann, Zacharius Brendel, Daniel Sennert, Andreas Libavius, Matthias Untzer, Heinrich Petraeus, Bernard Georges Penotus, Angelo Sala, Oswald Croll, Johann Daniel Mylius, and Johannes Rhenanus (Reinesius [1624] 1678: 3, 25–6). The list was composed too early to mention the Brussels-born physician and chemist Jan Baptiste van Helmont, one of the founders of another tradition of chemical medicine called iatrochemistry. For van Helmont chemistry did not just offer the means to making good medicines; it provided the key to understanding nature. "I praise a generous God who called me to the art of the fire ... For, more than all the other sciences, *chymia* prepares the intellect for penetrating to the hidden parts of nature, and thus penetrates to the furthest depths of objective truth" (van Helmont 1667: 286).

Many people agreed that in some fashion chemistry deserved a place as part of university education. In 1609 the preparation of chemical medicines briefly became part of the curriculum at the University of Marburg. At the University of Jena, Zacharius Brendel Jr insisted upon its presence as essential to the study of medicine, and proclaimed that those who wanted to remove chemistry from medicine were like those who wanted to remove the sun from the world. Later, Robert Boyle insisted on much the same thing, describing chemistry as the very backbone of medicine since it extracts the more active parts of bodies and enriches the virtues of remedies. Neither medicine nor the chemical art ruled supreme. In fact, each was so dependent upon the other as to be inseparable. "Let us unite medicine and *chymia*," Brendel proclaimed, "each one as master and slave to the other" (Moran 2015: 74–6). Elsewhere in Europe the study of the chemical art, aided by early textbooks, took further root in academic settings as recognition grew of its role in perfecting other arts and sciences and of its capacity to teach what was fundamental to the construction of the world (Debus 1990).

CURIOSITY AND CHEMICAL PLACES

The intersection of chemical theory and practice with educational, economic, and intellectual realms of experience not only equipped the early modern era with new products and new debates, but also played a role in fashioning its perceptions, desires, and experiences. Among the cross-currents of social and intellectual life, wonder, utility, and, in certain instances, playfulness combined as powerful elements of curiosity. Yet curiosity itself was not a stagnant concept, and studies of its historical mutability stress a change in the view of curiosity, beginning in the mid-sixteenth century, which refashioned personal inquisitiveness from a vice to a virtue, and finally into an epistemological norm. When separated from moral strictures, curiosity helped create a "new sensibility" in the cultural life of Europe that oriented attention toward possessing nature's secrets and duplicating her artistry. Curiosity emphasized the novel, the extravagant, and the rare, and, by the seventeenth century, a desire for knowledge (Daston 1995; Kenny 1998; Evans and Marr 2006). Sometimes "curious knowledge" meant secretive knowledge, the kind of knowledge known to only a few people, including chemists. What was curious could also refer to the craft involved in making things, the skills of material creativity. Among the trades listed by the English diarist John Evelyn in 1660–1661 on behalf of the "Philosophic Club" that was soon to become the Royal Society were occupations listed in eight categories, including "useful and purely mechanic," "polite and more liberall," and "exotick and very rare seacretts." One category labeled "Curious" included enameling, painting, printing, and engraving, and referred to the makers of clocks and all automata, spectacles, and mathematical instruments. It also included those involved in "alchimy," and specifically to the making of "bay salt, white salt, alum, coprose [copperas, i.e. ferrous sulfate], saltpeater, sulfur, etc." (Forbes Sieveking 1923–4: 46).

Some forms of curiosity centered upon natural wonders. Others reflected an attraction to nature's delicate workmanship. Both had long come together in collections of objects, some natural, some made by human craft and artisanry, in what were called cabinets of curiosity. The collections of private, usually wealthy, collectors mirrored personal passions. Many took shape at notable European courts. At Dresden, the confluence of artisanal expertise, tools, mathematical instruments, and uncommon material objects in the collection of the Elector August of Saxony gave rise to a site that emphasized trades and crafts (Bäumel 2004). The demonstration of artisanal know-how was also a prominent feature of collections at the Medici court. At the Casino di San Marco, Francesco I de Medici first set up a *Fonderia*, a laboratory/workshop for things made by fire. Thereafter, laboratories and artist workshops appeared also at the Uffizi, and they were administratively brought together by Francesco's successor, Ferdinand I, as the *galleria dei lavori*. The *Fonderia* produced a wide variety of metallurgical, chemical, and pharmaceutical products that combined craft techniques with court desires. Making artificial precious stones, fireworks, explosives, imitation Chinese porcelain, poisons, and antidotes combined there with projects in painting, sculpture, and goldsmithing. Alchemists and painters frequently collaborated in turning works of nature into works of art (Kieffer 2014), and as the workshop grew it required a supervisor of artistic productions to oversee all "jewelers, carvers of any type, cosmographers, goldsmiths, the makers of miniatures, gardeners of the gallery, turners, confectioners, clockmakers, artisans of porcelain, distillers, sculptors, painters, and makers of artificial gems" (Butters 2000: 144). In one project especially the relationship between metalworkers and alchemists fit precisely into the symbolic expression of power at the Medici court as alchemists developed recipes for hardening metallic cutting tools that made possible working on *pietre dure* like porphyry, a technique that proclaimed the power of a dynasty to shape and control even the most resistant natures (Butters 2000).

Beyond the court, although sometimes with aristocratic support, collecting led as well to a new form of social organization and scientific community. Interest in exploring the ambiguities of nature became the focus of the Italian aristocrat Federico Cesi, who pursued projects that redirected attention away from wonders and marvels and toward collecting examples of "middle natures," those parts of the natural world that challenged the regularity of things by being partly one thing and partly another. In 1603, Cesi, along with three friends, founded an informal fellowship called the *Accademia dei Lincei* (Academy of the Lynx) aimed at the study of nature by means of observation, experiment, and inductive reasoning. Federico and his father had different ideas about what brought honor to an aristocratic lineage, however, and for a short time Federico was forbidden to associate with the academy's early members. Nevertheless, Cesi's own interests could not be suppressed. The Academy expanded, although never numbering more than twenty members, and included the Neapolitan polymath Giambattista della Porta, who concerned himself, among other things, with distillation and transmutation, and, most prominently, Galileo (Freedberg 2002: 65–77, 322–30). Another early scientific academy, the *Accademia del Cimento* (Academy of Experiment, established 1657) depended directly upon the organization and financial support of Leopoldo de' Medici.

Curiosity also became a key feature of an early German academy devoted to the study of nature, the *Academia* or *Collegium Naturae Curiosorum* (*Academy of Those Curious about Nature*), founded in 1652. Later, the academy became known as the Leopoldina when it added experiment to curiosity as part of its official program. Although initially a society of physicians interested in the medicinal properties of various parts of the physical world, its members soon demonstrated abilities in a variety of alchemical procedures related to the manipulation of acids, minerals, and metallic salts, with a focus in some instances upon transmutation. For members of the Leopoldina *chymia* thus became a subset of alchemy and involved some of its members in the defense and laboratory pursuit of transmutation, as well as in the refinement of useful recipes and practices related to *chymiatria* (Garber 2015).

Some associations were more informal and emerged as a result of personal friendships. In this way the early seventeenth-century French scholar and patron Nicolas-Claude Fabri de Peiresc helped advance the prospects of specific individuals, including Tommaso Campanella (an advocate of empirical science and a defender of Galileo), the mathematician Marin Mersenne, and Pierre Gassendi (experimentalist and proponent of an atomist theory of matter). The circle fashioned by the German–English "intelligencer" Samuel Hartlib is arguably the most famous and influential of such associations in the early modern era. Based in a company of personal friends, a cluster of companions that included Robert Boyle, Hartlib's network (his surviving correspondence and notes run to 25,000 pages) created a circle of contacts throughout Europe and reaching to America. Among those exchanging information related to alchemy, chemistry, and minerology were, besides Boyle, the "chymical gentleman" Cheney Culpeper, the mathematician and alchemical enthusiast Ezekiel Foxcraft, the Paracelsian alchemist, translator of alchemical texts, and author of the widely read *Art of Distillation* (based on the work of Johann Rudolf Glauber) John French, and the chemical geologist, metallurgist, and author of *Caveat for Alchymists* Gabriel Plattes.

Both Hartlib and Boyle took part in early discussions aimed at forming a scientific society in England. By the end of the seventeenth century the English Royal Society and the French *Académie Royale des Sciences* emerged as new kinds of institutions, with new forms of communication, oriented toward the collective and systematic exploration of the natural world. Each maintained its own methodological preferences and developed its own means of reporting conclusions and discoveries through official publications. The Paris *Académie* (founded in 1666) was financed by the state, receiving as a consequence not only an annual budget, but also a list of projects to be carried out by its members related to the state's interest in commerce and trade. The Royal Society, on the other hand, remained relatively impoverished even with its royal charter of 1662, but benefited from a freer hand in selecting projects. The result was that the Society soon settled into the likeness of a gentleman's club, but one nevertheless oriented toward ventures based in experimental learning. Both societies collected and published findings in what might pass for the first scientific journals, the French *Journal des Sçavans* and the British *Philosophical Transactions*, and both included considerations and debates related to a wide variety of chemical subjects, among them discussions concerning the corpuscular theory of matter and examinations of the experiments of one of the Royal Society's most prominent members, Robert Boyle.

EXPERIMENT, METHOD, AND "NEW LEARNING"

Shortly after the founding of the Royal Society, Boyle engaged in a debate, mediated by the future secretary of the Society, Henry Oldenburg, with the Dutch philosopher Baruch Spinoza. The dispute had to do with the necessary conditions of proof. Spinoza insisted on the role of logic. To Boyle, however, something else was necessary, namely experiment. Without experiment, Boyle argued, logic could never get out of its own way, demonstrations being based upon *a priori* assumptions. Experiment, however, put assumptions, and their supporting logic, to the test. In articulating the idea of experiment, the English philosopher and statesman Francis Bacon, and his influential text *Novum Organum* (1620), was of crucial intellectual significance; although, in many ways, alchemists and chymists had long before, in their own practices, been employing a key ingredient of the Baconian method. To learn useful things about nature, it was necessary to make things happen; that is, to force nature to do things that she would not do on her own. In his text Bacon argued that in learning nature's secrets one had to put nature on the rack, so to speak, and wring out the truth that otherwise remained hidden. Real learning occurred not when nature was "free and at large," but when nature was "under constraint and vexed; that is to say when by art and the hand of man she is forced out of her natural state, and squeezed and moulded" (Bacon 1960: 25). Furthermore, one had to follow a particular method of inquiry, the method of true induction, in which theories arose as a result of collecting and organizing individual observations and natural facts. Most important, Bacon recognized what has been called the pathology of human thinking, the notion that the mind, unless itself constrained, in this case by method, is prone to bias. "For the mind of man," he wrote, "is farre from the Nature of a clear and equall glasse, wherein the beams of things should reflect according to their true incidence; Nay, it is rather like an inchanted glasse, full of superstition and imposture" (Bacon 1605: 55v). Even sense perception fell victim to preference. "It is a false assertion," he observed, "that the sense of man is the measure of things. On the contrary, all perceptions as well of the sense as of the mind are according to the measure of the individual and not according to the measure of the universe" (Bacon 1960: 48).

Bacon imagined an organized learned society, "the noblest foundation ... that ever was upon earth," whose members would collect and compare knowledge of "the sciences, arts, manufactures, and inventions of all the world" and find out "the true nature of all things whereby God might have more glory in the workmanship of them, and men the more fruit in their use of them" (Bacon 1659: 17). The grand vision of the expansion of learning by means of cooperative inquiry, experimental learning, and royal patronage influenced the encyclopedic vision of Hartlib and his circle, and oriented as well the early organization of the Royal Society. In this context natural philosophy, in the form of public knowledge, gained a civic function and a practical purpose. Experimental learning rooted in "industrious observations, grounded conclusions, and profitable inventions and discoveries" supported both the pursuit of natural knowledge as well as the extension and maintenance of political power (Martin 1992) (Figure 0.3).

Bacon was skeptical of received opinions grounded in ancient authorities, and in this he shared much in common with the French philosopher René Descartes. Descartes compared knowledge based in ancient opinion to ivy that could climb no higher than the tree supporting it and that even tended to grow downward again after it had reached the top. To make knowledge for oneself one had to be guided by rational method, and in Descartes' view method began with intuition and experience. In defining what we would today describe as the deductive method he argued that experience and "certain germs of truth" that were naturally in the mind led to the formation of ideas, or hypotheses, and that these were further tested by experiment. But Descartes also divided the world of experience into two sorts of existence: things of the mind or thought (*res cogitans*) and those things that corresponded to the physical world and took up space (*res extensa*). The certainties of natural philosophy were limited to the latter; and in the investigation of nature, reference to two things only grounded physical explanation – matter and motion.

For chemistry the interrelationship between theory and epistemology did not have to wait until the Enlightenment. Descartes' notion – precisely stated in his *Discourse on Method* – that the road to truth was paved by an analytical method that reduced complex problems into their smallest components was already part of the rhetoric of chemistry long before Lavoisier. Analytical reasoning, when combined with mechanical thinking, produced new, mechanized theories of matter at the expense of traditional ways of understanding nature in terms of

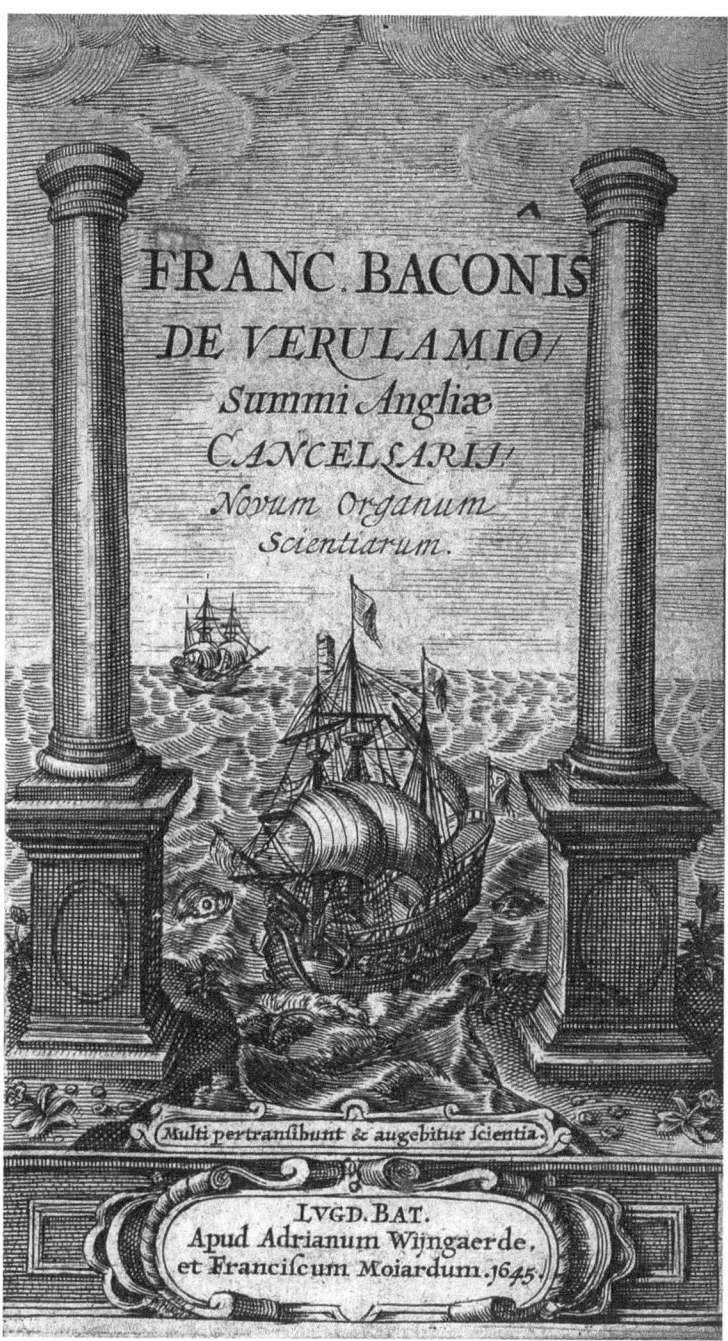

FIGURE 0.3 Frontispiece to Francis Bacon's *Instauratio magna* showing ships at sea sailing through the Pillars of Hercules to gain knowledge through exploration and by surpassing traditional learning. Credit: Wellcome Collection, London. CC BY.

inherent active powers. The seventeenth century was a world of practical objects useful for understanding mechanical properties and principles. The behavior of beams, pendulums, springs, and levers became increasingly relevant to natural and experimental philosophy as a developing science of mechanics replaced Aristotelian physics with its references to qualities and substantial forms (i.e. the innate sources of a body's properties).

For the cultural history of chemistry the attack on qualities and substantial forms, guided by new discoveries in the physical sciences, possessed significant meaning as it prompted a revival of thinking in regard to a view of matter composed of atoms. The notion was already widespread in the later sixteenth century, and one of the earliest chemical atomists, Daniel Sennert, resourcefully combined traditions of Aristotelian physics, ancient atomism, and the doctrine of substantial forms (now linked to the properties of individual atoms) in explaining chemical operations by means of particles. It was, however, the seventeenth century that witnessed the merger of atomist or corpuscular matter theory and the mechanical philosophy as a way of explaining, by means of constituent particles, the properties of material bodies. In describing the corpuscles or particles of air, Robert Boyle explained that "these are as it were the Seeds or immediate Principles of many sorts of Natural Bodies," which although "singly insensible, become capable, when united, to affect the Sense." Multitudes of corpuscles created clusters of particles, and the combination and dissolution of those particles became the basis for the generation, corruption, and alteration of physical bodies. Corpuscles, therefore, were the physical building blocks of bodies. Each had its own "Bulk and Shape," and the physical properties of the bodies that they created depended upon their size, shape, "Juxta-position and Cohaesion," and, especially, upon their distinctive motions. Changes occurred when "the motion of either one or the other ... will receive a new Tendency, or be alter'd as to its Velocity, or otherwise ... If many or most [of these corpuscles] be put into Motion, from what cause soever the Motion proceeds, That itself may produce great Changes and new Qualities in the Body they compose." Compounds acquired new properties as "Moved parts strive to communicate their Motion ... to some parts that were before either at Rest or otherwise mov'd ... [and] either drive some of them quite out of the Body, and perhaps lodge themselves in their places, or else associate them anew with others" (Boyle 1999c: 5, 325–7). Chemistry, in Boyle's hands, held the key to physical theory.

It would make a tidy picture if classifications and traditions were neatly bounded, if in the early modern era Aristotle and Paracelsus simply gave way to experiment and atoms, if alchemy gave way to chemistry, and if a philosophy of matter and motion replaced magic and vitalism. That, however, would be a picture created by fantasy rather than lived historical experience. Europe's sixteenth and seventeenth centuries mixed natural philosophy with theology, both Christian and Rabbinic, and blurred categories that may today appear distinct. Traditions of magic, Christian Kabbalah, Neoplatonism, Paracelsianism, Aristotelianism, and corpuscularianism shaped the amorphous and ever shape-shifting idea of what counted as "chemistry." All of nature, wrote the physician, hermetic philosopher, and alchemist Heinrich Khunrath in 1595, proceeds from God "theosophically, physically, physico-medically, physico-chemically, physico-magically, hyperphysico-magically, [and] Kabbalistically" (Khunrath 1595: pl. 2; Walton 2011: 77). It may be tempting to view chemistry as distinct from heterodox and mystical beliefs, but that would be to remove the essence of what many thought "chemistry" to be. Like it or not, spiritual commitments were, and are, part of knowing. The spectacle and mysteriousness of nature was the dynamic that gave the experience of the physical world its real meaning in the early modern era and that allowed chemistry to become a vital element of the period's culture – brought to bear, as the following chapters demonstrate, in objects and their making, in the production and trade of commodities, in the fashioning of schools and institutions, as well as in the practices that relate to the extension of theoretical knowledge and to the designs and representations of the visual arts.

INTRODUCTION

CHAPTER ONE

Theory and Concepts: *Conceptual Foundations of Early Modern Chymical Thought and Practice*

LAWRENCE M. PRINCIPE

Chemistry was both widely practiced and highly visible during the sixteenth and seventeenth centuries, as well as extremely diverse in terms of its theories, concepts, and goals. Indeed, this time period marks a rapid expansion of the practice and importance of chemistry generally. Present in various forms and with various goals from one end of the social spectrum to the other, chemistry provided a major – if still underappreciated – locus for human engagement with the natural world, endeavors to understand it, and practices to direct and harness it toward a range of useful and profitable practical ends. Long established as a subject that linked both head and hand, theory and practice, chemistry blossomed in the early modern period. It expanded into new realms and diversified in multiple ways. This burgeoning of the discipline yielded an abundance of new theories to explain observations and to guide practices, and these newer concepts often coexisted alongside older and longer-established ones.

The extraordinary range of theoretical and conceptual principles present in early modern chemistry resulted from two major factors that confronted every serious practitioner. The first is the inherent ontological richness of chemical phenomena. Chemistry is primarily concerned with the *qualities* (e.g. colors, smells, tastes, textures) and *properties* (e.g. solubility, corrosivity, volatility) of material substances and how such materials transform and interact with one another. The sheer range and variety of qualities and properties, as exhibited by an ever-increasing roster of known chemical substances, and the giddy and seemingly unpredictable ways in which such qualities and properties change through chemical processes, presented practitioners with a riotous chaos of experiences and observations. Many chemists thus naturally sought more general theoretical principles that could organize this observational chaos and provide some degree of both explanatory clarity and operational guidance. But a second factor complicated this organizational aspiration: the fundamental principles that underlie the nature of matter lie beyond the reach of human sense perception. How then can one fundamentally explain or predict the appearance and disappearance of various qualities and properties, if the deepest levels we can actually sense are those qualities and properties themselves? Given the inherent difficulty of the task it is no surprise that so many different explanatory schemes were devised in the early modern period, and it becomes all the more remarkable in hindsight that so much of lasting value was actually accomplished at the time. When surveying the attempts of early modern practitioners to make sense of what they were witnessing in their crucibles, retorts, and alembics, it is important to bear in mind these twin difficulties that nature itself constantly presents to the chemist.

Before progressing further, it is necessary to establish a consistent and historically sensitive nomenclature for describing the endeavors of the period. The early modern historical actors used many different terms to describe their own practices and goals. The two general terms for their discipline that persist most widely with us today are *alchemy* and *chemistry*. Since the early eighteenth century, those two words have generally held significantly different meanings. The first suggests something archaic, mysterious, and of questionable legitimacy. The second remains attached today to the most important discipline of the physical sciences. Nevertheless, before the early eighteenth century the two words were used largely interchangeably or at least did not carry the connotative baggage that they do today. Both words stem from the same Greek term, *chemeia* (probably based upon *cheō*, meaning *to melt*). The term came into the Latin lexicon in the Middle Ages as *alchemia*, bearing the Arabic definite article *al-* as a prefix, a signal of the subject's transmission to Europe through Arabic sources. The subsequent orthographic bifurcation between *alchemia* and *chemia* originated in attempts by Renaissance humanists to "purify" the Latin language from nonclassical accretions by removing the Arabic appendage. This endeavor had only limited success, meaning that both *alchemia* and *chemia* (and their vernacular cognates) coexisted for generations afterwards. This etymological digression is important for our present purposes because it would be highly misleading to attempt to divide up early modern theories, concepts, or practices into the *retrospective* categories of alchemy and chemistry. To address this issue, it has been proposed – and has already become common among scholars – to use the archaically spelled *chymistry* in order to refer comprehensively to the early modern subject and to avoid imposing anachronistic assumptions and value judgments upon our historical analyses (Newman and Principe 1998). I shall use the term henceforth in this chapter as appropriate.

Early moderns used a variety of other names to refer either to chymistry as a whole or to more specific subsets within it. The use of these terms where appropriate facilitates rigorous historical analysis. The important topic of metallic transmutation, a subject often now uncomplicatedly equated with the modern meaning of *alchemy*, was more specifically called *chrysopoeia* (from the Greek for *gold-making*) or less commonly *argyropoeia* (*silver-making*), and was predominantly seen as a subset of the broader *chymistry*. The equally important goals of making improved medicinal preparations through chymical manipulations and understanding the body through chymistry garnered their own terms, particularly *iatrochemistry* and *chemiatria*, both incorporating the Greek *iatros* (physician). For those pursuing pharmaceutical goals along specifically Paracelsian lines, there was also the term *spagyria*, which strictly speaking refers to a particular method of preparing medicinal substances chymically (see below).

Because chymistry, perhaps more than any other premodern subject that developed into one of our modern sciences, transcended any meaningful division between natural philosophical knowledge and material production, it occupied a peculiar cultural position straddling craft and science (*ars* and *scientia*). Insofar as it involved speculations, theorizations, and explanatory principles regarding the nature, properties, and transformations of material substances it ranked as a part of natural philosophy, and many chymists routinely referred to themselves as *philosophers*. The early seventeenth-century Flemish chymist and physician Jan Baptiste van Helmont labeled himself a "philosopher by fire," indicating that his ideas and understanding of nature came from his practical experiences in the fire of the chymical furnace rather than from mere cogitation or verbal wrangling. Insofar as chymistry involved the practical production of material substances, whether precious metals, medicines, or other objects of commerce, it was considered a craft or art. This dual nature of the subject expressed itself not only in references to its practitioners either as "philosophers" or as "artists," but also in an unstable and often contentious identity for chymistry that was not resolved during the early modern period, and that arguably remains incompletely resolved even within modern chemistry today. Chymistry's many early modern practitioners situated themselves everywhere along a broad continuum stretching between the highly speculative and the exclusively practical. While the balance of this chapter focuses on the theoretical and conceptual frameworks for early modern chymistry, the generally inseparable nature of the theoretical and the practical/productive within the subject must not be underestimated.

THEORIES OF COMPOSITION: PRINCIPLES AND PARTS

A major preoccupation for chymistry, both for fulfilling its explanatory aims and for guiding its practical endeavors, was the desire to develop coherent and useful theories of material composition, a topic today sometimes called "matter theory." Such theories fall into two overlapping categories. Theories in the first category endeavored to identify the basic or ultimate components of material substances. Today, one would be likely to call these ultimate components *elements*, but that word was used only in restricted instances during the early modern period; the simpler components of compound substances were more usually called its *principles*. This word implies that such substances are compositionally simpler than the bodies they compose, but without necessarily entailing that they are the ultimate and *most* simple components of compound substances. The second type of theories about composition endeavored to describe what matter is like at the submicroscopic level; for example, whether it is continuous (with no gaps or breaks) or particulate, and how stable those particles might be. Both sorts of theories underwent considerable development during the sixteenth and seventeenth centuries.

The early modern period inherited four major chymical matter theories from the Middle Ages, all of them stemming ultimately from Aristotle, although with substantial developments during the intervening two thousand years. These concepts underwent further changes in the early modern period as they were adapted to or refuted by new ideas and observations. The first of these matter theories was Aristotle's concept of *hylomorphism*: namely, that material bodies consist of two distinct entities, a quality-less *material* often called "prime matter" upon which is imposed a *form* (or multiple forms in some interpretations) that provides all the qualities of the body. For Aristotle, prime matter was a logical concept rather than something one could put into a bottle. It was not something that literally existed on its own but was instead a conceptual "part" of a substance. Another matter theory involved the idea of the four elements – fire, air, water, and earth – which were thought to be elemental or quasi-elemental building blocks of all material substances and to arise from a pair of the "primary qualities" (hot, cold, wet, dry) imposed upon prime matter. For those interested in material production (i.e. chymists), there was a range of opinions about how concrete these four elements actually were – were they actual isolable materials, logical constructs, prototypes for classes of physical substances, or something else? A third and more developed idea popular among chymists and particularly favored by those interested in *chrysopoeia* postulated two substances – mercury and sulfur – as the principal constituents of metals and minerals. These two might, or might not, be identical to or related to the common substances we know today by those names, depending upon the author in question. Although based ultimately upon the twin exhalations proposed in Aristotle's *Meteors* that arise from within the Earth and that generate stones below ground and weather phenomena above ground, the mercury–sulfur theory was first elaborated in Syriac or Arabic sources of the early Middle Ages (Principe 2013: 35–7). Finally, there also existed a quasi-particulate conception of the structure of matter that attributed chymical changes to the varying juxtapositions of the smallest parts of material substances, often called *minima naturalia*. This fourth theory of matter, deriving ultimately from suggestive passages in the fourth book of the *Meteors*, was promoted especially by the highly influential thirteenth-century Latin chymical author Geber, and so found a stable place within chymical speculations thereafter (Newman 2006a: 23–44; Calvet 2010).

Some of these concepts were better accommodated to guiding or explaining practical work while others found more use in philosophical speculations, yet they all coexisted (and frequently enough overlapped) in the early modern period. For example, while many authors continued to use hylomorphism for rather generalized explanations (e.g. that chymical changes require that an initial form be stripped from prime matter and a new form subsequently introduced), this provided little or no practical guidance for laboratory work. Nevertheless, early modern chymists, more focused on actual laboratory practice, tended to view prime matter in more material terms than Aristotle did. For them, if prime matter could be isolated, or at least generated practically, it would offer a material blank slate upon which any form might be imposed, thereby producing any desired substance. Rather more helpful to the practical chymist, however, and therefore more widely deployed, were the systems that posited the existence of more clearly *material* principles as the building blocks of chymical substances. The addition, subtraction, or modification of these principles in compound substances were thought to account for chymical change. Both the four-element theory and especially the mercury–sulfur theory of the metals fall into this category. It was the latter of these two that underwent the greatest degree of modification and development in the early modern period, and in its various forms was the more widely accepted and utilized.

A major development of the mercury–sulfur theory was proposed by the Swiss physician and iconoclast Theophrastus Bombastus von Hohenheim, called Paracelsus. Paracelsus added a third principle, salt, to mercury and sulfur, and extended this new triad – eventually called the *tria prima*, or "three first things" – to include all substances, not just the metals and minerals. The concept of the *tria prima* was widely adopted by the numerous followers of Paracelsus. The *tria prima* were not necessarily believed to be truly elemental substances (i.e. the ultimate constituents of matter), although some authors did view them in this way. In some conceptions at least, each of the *tria prima* might be composed of yet simpler materials such as the four elements. But they were considered to be at least *simpler* substances isolable from naturally occurring materials, even if they were not truly elemental ones. The *tria prima* were frequently considered *classes* of material substances that were separable by chymical methods from compound substances and that the chymist could identify by specific observable properties. Sulfurs were characterized by their flammability, color, and smell; mercuries predominantly by their volatility; and salts by their taste, water solubility, and crystalline character (Hooykaas 1935; Principe 1998: 37–40). Each of these components contributed its own properties to the composite materials formed from their combination. The classic demonstration of the separation of the *tria prima* was a burning twig – the flame shows the presence of the inflammable sulfur, the smoke is the volatile mercury, and the ashes, once extracted with water and evaporated, yield the salt. The persistence and utility of the *tria prima* as a theory of composition was based upon practical results from the destructive distillation of plant and animal substances. When distilled, most organic bodies emit a volatile and an oily fraction – the substance's mercury and sulfur, respectively. A soluble, crystallizable salt can then be isolated from the nonvolatile distillation residue, called the *caput mortuum* ("dead head"). For Paracelsus and some of his followers, however, the *tria prima* had wider connotations beyond the practical manipulations of chymistry. In cosmological schemes, this material trinity was drawn into analogy with the triune nature of human beings (body–soul–spirit) and with the Holy Trinity.

The sixteenth-century expansion by the Paracelsians of the medieval binary of mercury–sulfur into the triad of mercury, sulfur, and salt took a further step at the start of the seventeenth century with the postulation of a *pentad* of principles: mercury, sulfur, salt, water, and earth. This pentad is most often connected with the French chymists Joseph Du Chesne (himself a Paracelsian) and Etienne de Clave (Hooykaas 1937; Joly 2001). The addition of water and earth to the list of principles responded to the experimental observation that the decomposition of organic substances gave more than just three classes of materials. Destructive distillation also provided an insipid, watery material that could not be reasonably denominated as either a mercury or a sulfur, and was thus identified simply as water. The extraction of the water-soluble salt from the *caput mortuum* left behind an insoluble residue, and this was called the earth. It should be noted that multiple names were attached to several of these principles – mercury was sometimes called spirit, sulfur oil, and water phlegm. The advocates of the *tria prima* were not ignorant of these additional components of compound substances, but considered them useless (or at best passive) ingredients in the context of pharmaceutical activity. The valuable, active properties ("virtues") of a substance were considered to reside in the other three principles.

Early modern chymical pharmacy, or *chemiatria*, aimed at concentrating, enhancing, or correcting the medicinal powers of naturally occurring substances. Paracelsus and many of his followers considered that the toxicity or dangerous side effects of potentially curative substances were not essential to their true nature, but rather consequences of the postlapsarian state of the world. Just as the transgression of our first parents had tainted human nature with original sin, that same fall of man had polluted even the material substances of an originally perfect creation with poison. According to this concept, however, any material, regardless of how toxic its ordinary effects, could be transformed into a medicine, more powerful than the then more standard Galenic medicinal preparations.

Hence Paracelsian chymists endeavored to remove the poisonous qualities from substances like arsenic, antimony, mercury, and other metals and minerals by means of chymical processes. This process was often labeled with the German term *Scheidung* (i.e. division or separation). In the endeavor to "correct" materials into more medicinal forms, the idea of the *tria prima* that contained all the virtues of the composite was especially valuable. The hope was to separate a given material into its mercury, sulfur, and salt, purify each of them separately, and then recombine them into a reconstituted and purified material. This "glorified" substance could then manifest its curative properties more powerfully, without having its properties suppressed or corrupted by the presence of useless or toxic impurities. This process was known as *spagyria*, a Paracelsian neologism supposedly coined from the Greek words for *to separate* and *to combine* (Principe 2013: 128–30). The emphasis on *spagyria* led to an enhanced importance for the processes of *analysis* and *synthesis* in early modern chymistry, and indeed, these reciprocal processes remain central aspects of even modern chemistry.

It is important to stress that the binary, the triad, and the pentad of principles coexisted throughout the seventeenth century. Practitioners chose the system that correlated best with the particular endeavors in which they were involved. For those working mostly with metals and minerals, predominantly chrysopoeians, some version of the mercury–sulfur theory was often quite adequate. The triad and pentad found more use among those dealing with plant and animal substances, particularly for the preparation of pharmaceuticals. The *tria prima* underwent further modification at the hands of the German chymist and cameralist Johann Joachim Becher in the second half of the seventeenth century. Unlike many proponents of the *tria prima* and pentad, Becher was interested predominantly in inorganic (metallic and mineral) substances. While he rejected the *tria prima*, Becher still maintained the notion of a threefold division for the principles of compound bodies, but he renamed these principles as three types of *earth*: fluid earth, inflammable or fatty earth, and vitrifiable earth. These three are clearly alignable with the more traditional principles of mercury, sulfur, and salt, respectively. Water also exists in Becher's system as an additional principle that is able to combine with any of the three earths to produce composite materials. These initial composites can combine with each other to produce more complex combinations, and these latter then go on to combine yet further to form still more complex materials, thereby creating a hierarchy of increasingly complex composition – a concept also found in the works of Robert Boyle (Partington 1961: 2:637–52; Newman 2014). Georg Ernst Stahl revived and modified Becher's system at the turn of the eighteenth century, emphasizing especially the role of Becher's inflammable earth in chymical processes. Through Stahl's work, this inflammable principle became known as *phlogiston*, a term Becher himself occasionally used and that is derived from the Greek word *phlox*, meaning fire (Chang 2015). This phlogiston was liberated as a material burned – and it became the centerpiece of one of the most widespread explanatory systems of chymistry in the eighteenth century.

Some chymists of the seventeenth century, however, wished to pursue the identification of compositional substances to a degree of simplicity beyond that of the principles; that is, they wished to identify a *single* material foundation for all substances. The postulation of only a single, universal primordial substance (rather than of multiple elements) is known as *monism*. A highly influential theory along these lines was proposed by one of the seventeenth century's most important chymists, the Flemish natural philosopher Jan Baptiste van Helmont. Van Helmont combined chymistry, medicine, theology, experiment, and practical experience into a cohesive and highly influential system. He doubted that the materials separated by fire analysis that the Paracelsians called the *tria prima* actually preexisted in the composite substance being analyzed. The heat of the fire, he claimed, altered the decomposing compound and produced the separated portions rather than merely disentangling and isolating them (Debus 1967). He argued instead that the ultimate and sole fundamental substratum of all material substances is water. This concept resembles an idea connected with the ancient Greek philosopher Thales of Miletus: namely, that "everything is water." But for van Helmont, this theory gained special support from the mention in Genesis 1 of the "waters" upon which the Spirit of God brooded to create the world. Van Helmont's theory depended upon the action of immaterial *semina* (seeds or seminal principles) that are able to transform water into all the substances in the world. These *semina* act as organizing principles capable of inducing radical transformations of material properties – like the invisible principle that organizes the fluid of an egg yolk into a chick.

A keen experimenter, van Helmont performed multiple experiments to support his theory of the primacy of water. In the most famous of these experiments, van Helmont filled a large pot with two hundred pounds of dried soil and planted a young willow sapling weighing five pounds in it. He then let the willow grow for five years, watering it as necessary with pure rain or distilled water. At the end of this time, he uprooted the tree, scraped off the soil, and found that the tree weighed 169 pounds and 3 ounces. He dried the soil and weighed it again, and found that it weighed only two ounces less than when he had first planted the tree. Thus van Helmont concluded that the 164 pounds of increased weight of the tree (not counting the weight of the leaves that had fallen off during four fall seasons) must have been produced from water alone. All the different substances contained within the willow – the oils (sulfurs), spirits (mercuries), salts, and earths that could be separated from its wood by distillation – had been generated out of water alone through the actions of the organizational *semina* within the growing tree (van Helmont 1966: 71, 108–9; Pagel 1982b: 49–60; Hirai 2005: 439–62).

According further to van Helmont's theory, the materials formed by the action of *semina* upon water return to water when the *semina* are destroyed. Burning and putrefaction in particular weaken the *semina* and their organizational power and thereby turn substances into air-like materials that van Helmont called "Gas," substances whose *semina* have been mostly but not entirely destroyed. Burning charcoal and fermenting wine release a choking *Gas sylvestris*, and burning sulfur a stinking *Gas sulphuris*. Other sorts of *Gas* are produced during digestion in the stomach and by the action of acids upon metals. Van Helmont coined the term *Gas* from the Greek word *chaos*. Van Helmont's term was revived by eighteenth-century chemists, for whom the study of gases was important, and was used thereafter with its modern meaning. One of van Helmont's experiments showed him that when sixty-two pounds of charcoal are burned, they leave behind only one pound of ash; therefore, he reasoned, sixty-one pounds must be converted into *Gas* that escapes into the air. The various sorts of *Gas* rise into the colder parts of the atmosphere where their retrogradation back into primordial water is completed. The resultant water falls to the ground as rain (where *semina* can transform it anew into other substances), thus closing the cycle of water's successive transformations (van Helmont 1966: 72–81, 106–11; Pagel 1982b: 60–7). This retrogradation to water can also be accomplished in the laboratory, if one knows the right secret. Van Helmont claimed to have produced a powerful solvent that he called the *alkahest*. This fluid was supposedly able to break down any substance first into its proximate principles and then (upon heating and prolonged digestion) back into its primordial water by "mortifying" the seminal principles that had originally transformed the water. For the remainder of the seventeenth century, many chymists sought to prepare the alkahest as a means of carrying out a rigorous analysis of compound substances into their proximate principles in order to determine chymical composition and for preparing powerful medicines (Le Pelletier 1706; Joly 1996).

Monism was also a feature of many *corpuscularian* theories of the seventeenth century. Such theories claimed that matter is made up of minute, submicroscopic particles that possess various shapes and motions. According to some systems, such as that proposed in the 1640s by the French priest Pierre Gassendi, who endeavored to revive and Christianize ancient Greek atomic theories expounded by Epicurus, these particles moved freely in void space. In others, such as that advanced by René Descartes at about the same time, the variously sized particles packed together without leaving any interstitial voids (Osler 2010: 77–93; Joly 2011). Such particulate or corpuscularian thinking represented one important part of the so-called *mechanical philosophy*, a comprehensive theory of nature that endeavored to explain all natural phenomena mechanically in terms of matter and motion. In being applied to chymistry, the mechanical philosophy endeavored to reduce the sensible qualities of substances – which according to Aristotelians are due to their substantial forms – to the status of *secondary* qualities. That is to say, such secondary qualities do not actually exist, but are merely consequences of the ways in which the *primary* qualities – in this case, the shape and motion of submicroscopic particles, the only true qualities inherent in matter – act either directly or indirectly upon our senses. The yellow color of gold, for example, does not have any real existence in the metal, but arises only out of the way in which the texture of gold's agglomerated particles alter the light that the metal reflects and the way our eyes perceive that altered light. According to the mechanical view, hot and cold – which for Aristotle are two of the four primary qualities of objects – exist as such only in the senses rather than in a hot or cold object itself, for they arise from the faster or slower motion of the object's corpuscles and the way that motion affects the senses. Fluid substances are composed of particles that roll or slide upon one another easily; solid ones have their corpuscles tightly locked together. Early modern ideas about submicroscopic chymical particles rarely consider them to be indivisible, and in this way they differ from most ancient ideas of atoms, which were defined as indivisible entities.

The application of corpuscularian principles to chymistry is most frequently associated with Robert Boyle and his publications in the second half of the seventeenth century. According to Boyle, all substances are composed of corpuscles made out of a single "Catholick" (universal) matter. The observable differences between materials arise from how the corpuscles are shaped, how they are agglomerated, and how they move. Hence, chymical transformations are the result of alterations in the shape, size, agglomeration, or motion of the corpuscles that constitute a substance. Boyle's idea was that the constituent parts of a compound could be separated and then put back together mechanically like the parts of a clock or other mechanical device. In a famous experiment he carried out in the mid-1650s upon saltpeter (potassium nitrate), he showed how this compound could be divided by chymical operations into a volatile and a nonvolatile part, each of which manifested its own particular properties differing from those of the original saltpeter. Boyle then recombined these two separated components and was able to "reintegrate" the original saltpeter with all of its characteristic properties (Boyle [1661] 1999a). Nevertheless, it is also clear that Boyle, who was deeply influenced by van Helmont and other predecessors in the chymical tradition, did not hold a strictly mechanical interpretation of corpuscularianism. He combined notions of *semina* and active principles with the more mechanical aspects of corpuscularian theories to provide a system that to his mind better explained the properties of matter. Indeed, Boyle's initial enthusiasm for mechanical explanations seems to have waned in his later years as he considered the shortcomings of the system for explaining observations and the unacceptable theological consequences of the deterministic universe that a strictly mechanical view of nature would imply (Clericuzio 2000: 106–48; Osler 2010: 126–30).

Mechanical chymical explanations were devised not only for static properties but also for chymical reactivity. One contemporaneous theory – originating ultimately from the writings of van Helmont – posited that chymical reactivity was due to interactions between the opposing properties of acids and alkalis. The more mechanical reinterpretation of this theory, developed significantly by the German apothecary (living in Venice) Otto Tachenius and elaborated particularly by writers interested in the chymical preparation of medicines, such as the physician François de Saint-André and the apothecary Nicolas Lemery, proposed that acids are composed of sharp, pointed corpuscles in rapid motion. Hence, such acid particles "prick" the tongue, giving the sensation of sourness, and due to their shape and motion are able, like tiny wedges, to break apart solid bodies mechanically into their smallest parts, thereby dissolving them. Alkalis, on the other hand, were envisioned as composed of corpuscles that are porous, like little sponges, their empty spaces capable of receiving and holding the pointy acid particles (Clericuzio 2000: 173–7; Principe 2007: 4–6). When mixed together, acids and alkalis combine to form a new substance with properties completely different from those of either the acid or the alkali. These new properties stem from the new shape of the particles produced from the mechanical combination of the spongy alkali and the pointy acid.

Despite the overt opposition to Aristotelianism voiced by many advocates of corpuscularianism, the concept was nevertheless based partly in the Aristotelian tradition and even recapitulated some of its ideas. For example, corpuscularianism's monist underpinning restored the notion of a single, primordial, and quality-less prime matter under the guise of the universal matter of which all corpuscles are made. Like prime matter, this universal matter has properties only to the extent to which it is shaped and moved; that is, only to the extent it is given a *form*, although this latter sort of form is inherently more physical and concrete than the more purely Aristotelian concept of form as a logical construct. In this sense one might say that the corpuscularian hypothesis resembles a modified hylomorphism transferred from the macroscopic world of sense perception to the invisible microscopic world. Furthermore, it is certainly the case that the chymical corpuscularianism of the seventeenth century stemmed in part from, or at least paralleled, an older theory of matter whose roots lie within the Aristotelian corpus: namely, the concept of matter's "parts and pores" and their utility for explaining natural phenomena, as found in the fourth book of the *Meteors*. Such ideas, expanded upon during the Middle Ages and applied to chymical phenomenon, accompanied chymical practices down to the seventeenth century, in large part because this system proved useful in rationalizing practical experimental results. For example, this concept referred the various densities of different materials to their relative degrees of porosity; substances with more pores were less dense than those with fewer.

Practicing chymists had also observed that the inherent identity of some materials was more robust and resilient than they might have expected. A metal could be dissolved in acids, turned into a salt, roasted into a powder, and put through yet further transformations, and yet the clever chymist could recover the metal in its original form through subsequent operations. Such processes, which came to be known as "reductions to the pristine state" (*reductiones ad pristinum statum*), implied that some substances were composed of robust particles that could survive such significant changes of appearance while remaining unaltered at the microscopic level, and thus they were able to reemerge as the starting material at the end of the process. Most forms of hylomorphism could not explain this result: how could the initial form, annihilated at the start of the sequence, suddenly be made to reappear at its end? Such processes were pursued and studied in the early seventeenth century by the German university professor and chymist Daniel Sennert (Michael 1997; Newman 2006a: 85–125; Klein 2014b). In a typical example, he dissolved silver in acid to form a transparent liquid, precipitated the solution with common table salt to provide a white powder, and then heated the white powder strongly with salt of tartar (potassium carbonate), which gave back the silver in its original weight with all its properties intact, as if nothing had ever happened to it. The silver must therefore have been separable into particles so small they were invisible in a solution and able to pass through the minute pores in filter paper, yet all that time retaining their fundamental identity as silver. The chymical corpuscularianism of the later seventeenth century drew upon such experiments – Boyle himself carried out probatory experiments that paralleled the earlier *reductiones ad pristinum statum* – as evidence that the identities of materials actually lie in the minute corpuscles of which they are composed.

While corpuscularian concepts did provide some plausible explanations for observed phenomena, the larger project of reducing chymistry to mechanical principles did not succeed. By the end of the seventeenth century, the enterprise had largely evaporated. Practicing chymists recognized that the scope and variety of chymical phenomena was simply too broad and too diverse to be accommodated by the lean and limited principles of mechanical action. Although *ad hoc* and *post hoc* rationalizations could be made for many chymical processes, mechanical explanations failed to provide a comprehensive understanding of the diversity of chemical reactivity and its specificity. If acids worked simply mechanically like wedges to corrode solid bodies, why did one acid (*aqua regia*) dissolve gold but not silver, while another acid (*aqua fortis*, today's nitric acid) could dissolve silver but not gold? Copper dissolves in acid to form a perfectly stable blue solution, but if a piece of iron is introduced, the copper precipitates out and the iron dissolves. What mechanical explanation could explain the apparent "preference" of the acid to combine with iron over copper? These kinds of questions would provide central problems and ample room for speculation during the eighteenth century as natural philosophers sought more successful and satisfying explanatory principles.

TRANSMUTATIONAL CHYMISTRY

Early modern chymists recognized seven metals: gold, silver, copper, iron, tin, lead, and mercury. (By the end of the seventeenth century, chymists had also discovered zinc, bismuth, and possibly other metals, but these were not ordinarily classed with the seven classical metals because they lacked one or more of the properties shared by the others.) Two of these seven metals – gold and silver – they called "noble" on the basis of their resistance to fire and corrosion, their beauty, and their rarity. The other five they considered as base or "ignoble" metals. Crucially, metals were generally conceptualized as *compounds* composed of the same few ingredients in different proportions or degrees of quality, not as elements as we think of them today. Gold was considered to be the most perfect of the metals – its stability and resistance to corrosion resulted from its perfect proportion and tight combination of the principles mercury and sulfur in a very pure state. The metals tin and lead, on the other hand, contained various impurities as well as too much of the liquid principle mercury, as indicated by their low melting points. In contrast, iron and copper contained too much of the inflammable dry principle sulfur, hence these metals burn when their filings are dropped into a fire, and they are difficult to melt into a liquid. The compound nature of metals meant that a chymist should be able to separate them into their constituent ingredients, or simply alter their proportions or degrees of quality of their ingredients, and thereby turn one metal into another through chymical manipulations. Such potential transformation of one metal into another was known as *transmutation*.

Observations suggested that metallic transmutation was a naturally occurring process going on underground constantly but slowly. Miners and refiners noted that metals are rarely found singly; lead ores almost always contain some silver, and silver ores some gold. This common observation suggested that base metals naturally transformed underground into more noble ones as their compositions were slowly altered by the action of subterranean heat and water. Over hundreds or thousands of years, percolating groundwater slowly washed away the interfering impurities found in the base metals, while the gentle heat of the earth gradually cooked the baser metals into the better decocted, more stable, and perfectly united composition characteristic of noble metals. Therefore, the chrysopoeian needed only to find a way to do aboveground and quickly what nature was always doing underground and slowly.

Chymists interested in metallic transmutation – which means most of them to one degree or another during the early modern period – could pursue various practical pathways to that goal. In general, manually adjusting the proportions of a metal's ingredients was considered too difficult and time-consuming. Most chose instead to seek for an *agent* of transmutation, a material preparable in the laboratory that, when combined with a base metal, would bring about transmutation. These putative agents fell into two categories: particulars and the universal (Principe 2013: 112–14). The particulars, or *particularia*, were weaker transmutational agents that could transform only particular base metals into silver or gold. Thus, one particular could turn copper into silver, for example, but would have no activity toward other metals. The implicit contrast here is with the *universal* transmuting agent, known since late antiquity as the philosophers' stone, and somewhat later as the elixir, a material able to transmute any base metal into a precious one. The philosophers' stone was reputed to have far greater potency than any particular, capable of transmuting hundreds, even thousands or tens of thousands of times its weight of a base metal into gold or silver.

Searching for a means of preparing the philosophers' stone was an important part of early modern chymistry. Naturally enough, given the enormous power promised by the philosophers' stone, its method of preparation had to be kept secret. Many authors claim to have prepared it successfully, and many witnesses (including Boyle, among many others) report having seen it turn lead and other base metals into gold or silver. Contrary to some notions encountered today, stone-seeking chymists generally did not mix up "witches' brews" of various ingredients more or less aimlessly in the hope that they might thus stumble upon the formula. Most chymists instead guided their quest and based their belief in the existence of a transmutatory agent upon coherent natural philosophical principles of the day. Also contrary to some claims made in the older secondary literature, and still found frequently today in "popular" treatments of "alchemy," the vast majority of early modern seekers after the stone considered it to be a chemical substance preparable in the laboratory that acted upon base metals by completely natural, physical means. The stone was *not* "magical." The physical, chemical nature of the philosophers' stone itself did not prevent some authors from deploying the idea of the stone as a rich metaphor for a range of nonphysical or spiritual transformations, nor from appreciating and exploring the analogical ramifications of the stone across other domains. The point here is that the primary chymical literature of the premodern period is clear that virtually all chrysopoeians aimed at producing a physical substance in their flasks and furnaces that would literally transmute base metals into gold (Principe and Newman 2001; Principe 2013: 188–206).

The key to this endeavor was to discover the identity of the material(s) from which to prepare the stone. A wide array of materials were explored for this purpose – mercury and other metals, antimony, salts, organic substances, and so forth. Most sources stipulate that once found and correctly prepared, this material – or materials – needed to be sealed in a flask called the *ovum philosophicum* (philosophical egg) and heated for a long period of time. After about a month the material should turn black – a stage known as *nigredo* – a sign of putrefaction and change. Thereafter, the material, under the influence of increasing heat, was supposed to manifest various colors before becoming white. This white stage was the completed white elixir, which, after some further operations, could transmute base metals into silver. Continued heating of the white stage brought about its gradual transformation into a deep red color, the red elixir or fully mature philosophers' stone, which could transmute base metals into gold. Transmutation was carried out by throwing a small amount of the completed white or red stone onto a large quantity of molten metal, usually lead, or hot mercury. This process was known as projection, from the Latin *projicere* meaning "to throw upon" (Principe 2013: 115–25).

Ideas about how the philosophers' stone acted were likewise based on contemporaneous theories of chymical composition and a range of experimental and quotidian observations. Virtually all chrysopoeians agreed that the stone's action was explicable by natural principles like the rest of chymical activity. Everyone knew that a little vinegar thrown into a barrel of wine would transform it all into vinegar, that a tiny amount of rennet could coagulate many gallons of milk into cheese, and that small piece of leavened dough kneaded into a far greater quantity of fresh dough would leaven the entire mass. Common observations such as these provided precedents for the philosophers' stone's ability to coagulate a thousand times its weight of mercury into gold. Some authors claimed that the stone functioned like an especially intense refiner's fire, burning out the impurities from base metals that prevented them from being gold. Others held that the stone might be endowed with an excess of the Aristotelian form of gold, and hence, when projected onto a base metal, the stone destroyed the old form and replaced it with the form of gold. (Here again one can see how some early modern chymists began giving the Aristotelian logical concept of form a degree of physical reality that made it into something like an actual chymical entity.) Still others claimed that the elixir is "plusquamperfect" – that is, it is composed of gold elevated high above its usual level of perfection. When mixed with an imperfect metal, the imperfection of the metal and the plusquamperfection of the stone average out to perfection (i.e. to gold). Still other transmutational authors asserted that the stone contained the *semina* of gold capable of reorganizing other metals into gold. Yet other concepts explained transmutation using fully corpuscularian terms.

It should be stressed that the firm belief in the reality of metallic transmutation was widespread throughout the sixteenth and seventeenth centuries, as was the active pursuit of means of effecting it, particularly through use of the philosophers' stone. Robert Boyle endeavored to prepare the stone throughout his adult life, and he wrote a treatise defending transmutation and the stone (Principe 1998: 223–309). Isaac Newton's intense interest in transmutational chymistry, his fervent reading and interpretation of chrysopoetic treatises, and his extensive laboratory work are now well known (Newman 2018). The roster of eminent figures of the early modern age involved in chrysopoetic pursuits has become long and diverse over the past forty years and continues to grow with continuing studies by historians of science. Toward the end of the seventeenth century the quest for metallic transmutation and the preparation of the philosophers' stone began to receive increased public scrutiny due in part to stories, current and past, of fraudulent chrysopoeians bent on duping people with false promises. Despite this increasingly poor public image, chrysopoeia continued to be pursued by serious and well-established chymists – although increasingly secretly – far into the eighteenth century and even beyond (Principe 2014a). Indeed, historians are only now coming to realize the extent of chrysopoeia's persistence for generations after 1700 as it was repeatedly buoyed (rather than refuted) by new theories, concepts, and discoveries.

CONCLUSIONS

It is clear from the foregoing that the early modern period possessed a rich store of chymical theories and concepts, some new and others modified from older forms in various ways. A diversity of theories and concepts about chymistry characterizes the early modern period. Consequently, it is difficult, if not simply misleading, to point to any one single theory of chymistry as the "main" chymical theory of the period. The sixteenth and seventeenth centuries were times of both continuity and change, as chymical thinkers reworked the theories they had inherited into new forms or devised and proposed new systems and concepts to replace or complement them. It is therefore not correct to imagine – as has sometimes been done in the past – that sixteenth- and seventeenth-century chymistry did not possess sophisticated explanatory and organizational concepts and consequently had to await later developments (such as the work of Antoine Laurent Lavoisier in the late eighteenth century) in order to establish itself as a discipline. The fervent practical and theoretical work of a wide range of early modern chymists provided a rich and variegated identity for chymistry as a discipline and pursuit, even if the exact contours of that identity and a unifying theory remained to be synthesized.

CHAPTER TWO

Practice and Experiment: *Cultures of Chymical Analysis*

JOEL A. KLEIN

The early modern period witnessed transformational changes in how Europeans understood and interacted with the natural world. More people than ever before took up the study of nature and, aided by new technologies, they spawned and spread ideas at an unprecedented pace, altering the social, intellectual, and cultural landscape in ways that are still being felt on a global scale today. Indeed, historical figures living in the sixteenth and seventeenth centuries believed that something new and exciting was happening in their time, and many were eager to celebrate these changes. Authors announced the creation of "New Sciences" and "a new method of philosophizing," while new scientific societies established mottoes championing the value of experience. The Royal Society of London's motto *nullius in verba*, or "on the word of no one," perhaps distilled the essential tenet of the era: that experience was to be the primary arbiter for studying and knowing the natural world. This emphasis on practice and firsthand observation, the imperative to see, touch, or even taste something with one's own senses and not merely rest on authority, was codified within new experimental cultures where novel methods and systems of knowledge production would overturn or otherwise challenge many long-held beliefs about the natural world.

There were, however, many different and rival conceptions of what constituted a proper experience or experiment such that it would yield reliable information, and notwithstanding the rhetoric of many paragons of the New Science, there were significant continuities between medieval and early modern science. Far from nullifying the accomplishments of the early modern era, contemplating the complex historical lattice of conflict, continuity, and change is a main reason why the period remains so exciting for historians of science.

We now know that alchemy and early chemistry played a central part in the emergence of modern science. Perhaps more than any other area of natural investigation, alchemy deftly combined theory and practice, and it did so over its entire history. Even so, the history of alchemy is not marked by stagnation. Beginning in the Late Middle Ages, alchemy metamorphosed from an applied science focused on technologies and processes concerned chiefly with metals and minerals into an expansive philosophical and technical discipline that transformed medicine, led to new theories of matter, and had a major impact on the era's literary, artistic, and religious cultures.

But it has certainly not always been the case that alchemy and early chemistry were appreciated for their contributions to the history of science. From the Enlightenment up until the late twentieth century, the mainstream view among historians has been that alchemy was an unscientific, sometimes esoteric, generally foolish practice that was exclusively preoccupied with the philosophers' stone – the key to transmuting base metals into gold or silver (Principe and Newman 2001). Historians generally drew a strict line of demarcation between alchemical transmutation and other parts of the chemical enterprise that were viewed as being more congruent with modern science.

It has only been in the last several decades that this view of alchemy has been exposed as a caricature. Up until the late seventeenth century, most authors made no distinction between alchemy and chemistry, and while some natural philosophers and authors rejected the possibility of the transmutation of metals, this was not the prevalent opinion. Simply put, belief in the transmutation of metals was common and concordant with early modern chemical and intellectual culture.

A related difficulty in the historiography of early modern chemistry is that histories have traditionally focused on the development of chemical theory and have given less consideration to the practical side of chemistry (Klein and Ragland 2014). This focus on theory at the expense of practice is doubly problematic, for it not only obscures some of the most important developments in the history of chemistry, but it also erects an anachronistic barrier between theory and practice. Since its origins in the first centuries of the Common Era, alchemy was an applied philosophy that had both practical and theoretical components (Principe 2013). It was a textual and scholarly tradition, but at the same time incorporated artisanal and craft practice in workshops and laboratories, and whereas many other disciplines involved clear social, cultural, and intellectual distinctions between scholars and laborers, alchemy has always joined the head and the hand (Nummedal 2011).

This was true throughout the medieval era, when alchemy was understood partly as an artisanal pursuit that drew from natural philosophy and was concerned with material phenomena on earth and up to the boundary of the sublunary realm. Both Thomas Aquinas and Roger Bacon referred to alchemy as an "operative science" because of its practical ends and sensory means (Newman 1989). During the early modern period, alchemy underwent important changes and achieved much greater notoriety and cultural currency, but it continued to include a host of practical and artisanal components that it had evolved or incorporated over its long history. These included industrial and commercial activities such as the smelting of ore in mining, gold- and silver-smithing, diverse practices of apothecaries, the making of gunpowder and explosives, the refining and production of salts, and the fermentation of beer and wine. In addition, alchemy's domain touched on the decorative arts and included enhancements in glassmaking, the manufacture of dyes and pigments, and the production of artificial gemstones (Newman 2006b; Principe 2013).

Considering the coherence of transmutational alchemy with the rest of early modern chemistry, as well as its persistent multivalence as a practical and theoretical pursuit, scholars have recently begun using the archaic seventeenth-century term "chymistry" to signify both the unity and complexity of the alchemical/chemical enterprise while avoiding the modern connotation of alchemy as a pseudoscience (Newman 2006b: 499; Principe 2014a). The scholarly movement that has instigated this historiographical transformation, sometimes called "the new historiography of alchemy," has refocused attention on the unity of theory and practice in chymistry, but has also uncovered a great deal of historical complexity and problematized simple narratives (Principe 2011). Chymistry was understood in different ways by many different and sometimes adversarial groups, and it flourished amid a variety of contexts. The shift to focus on practice has considerably broadened chymistry's scope and has provoked new interest in exploring questions about who chymists were, what they did, and where they practiced (Nummedal 2011). Where did they source their materials and equipment, and how did they employ these in their practice? How did one go about learning chymistry? How did chymists communicate their findings, and what kind of cultures supported collaborative practice?

Scholars investigating these questions have turned to a variety of source materials beyond traditional books and textbooks, including recipe literature, laboratory notebooks, and letters. They have also utilized various sources of material and visual culture, from archaeological artifacts to paintings and printed images (Nummedal 2011). Some historians have even pursued historical questions by reworking or reconstructing historical experiments and processes in modern laboratories (Fors et al. 2016). In fact, this technique has proven to be an especially important resource for showing that early modern alchemy was a physical undertaking and that the allegorical and cryptic language of many alchemical texts was often a code for laboratory processes (Newman and Principe 2002). The use of these historical methodologies, combined with investigations of new sources, has made recent decades an especially exciting time for historians of this field.

Considering the rather extensive changes in our understanding of the scope and complexity of practices in early modern alchemy and chemistry, it is difficult to give a comprehensive account. The best that one can aspire to do is capture something of the essence of chymistry and hopefully avoid the pitfalls of generalization. I thus wish to focus here more narrowly on practices of chymical analysis and the related synthetic practices that involved by-products of analysis or the recombination of analyzed materials. These practices were central to several of the most significant chymical experiments carried out during the early modern era, and they allowed chymists to create new substances and develop new theories about the nature of matter. These endeavors were central to the rise of the experimental culture of chemistry and, in multiple respects, were at the heart of what it meant to be a chymist. Analysis and synthesis are likewise especially apposite for illustrating the interdependence of theory and practice, and their history cuts across important social and cultural boundaries during the early modern era.

ENDURING TECHNOLOGIES: ASSAYING AND DISTILLATION

Early modern chymistry's identity as an analytic and synthetic art containing elements of both theory and practice was in large part a holdover from medieval alchemy. This influence is especially evident in a body of works by a pseudonymous author who claimed to be the Arabic alchemist Jābir ibn Ḥayyān but was very likely a thirteenth-century Franciscan friar from southern Italy named Paul of Taranto (Newman 1991). Pseudo-Geber's corpus of alchemical works demonstrates in particularly sharp relief that the tradition of medieval alchemy bequeathed to early modern Europe and then propagated via print involved sophisticated and precise work in the laboratory (Newman 2006a) (Figure 2.1). Pseudo-Geber's *Summa perfectionis* (*Sum of Perfection*) was one especially influential text, and among its diverse and meticulously described processes were several assaying and analytic techniques that the author used to investigate and explain the fundamental nature of matter. These included cupellation and cementation – artisanal processes that had already been in use for centuries, but that were put to fresh philosophical use by medieval alchemical authors. Indeed, the discussion of these tests takes place in texts that

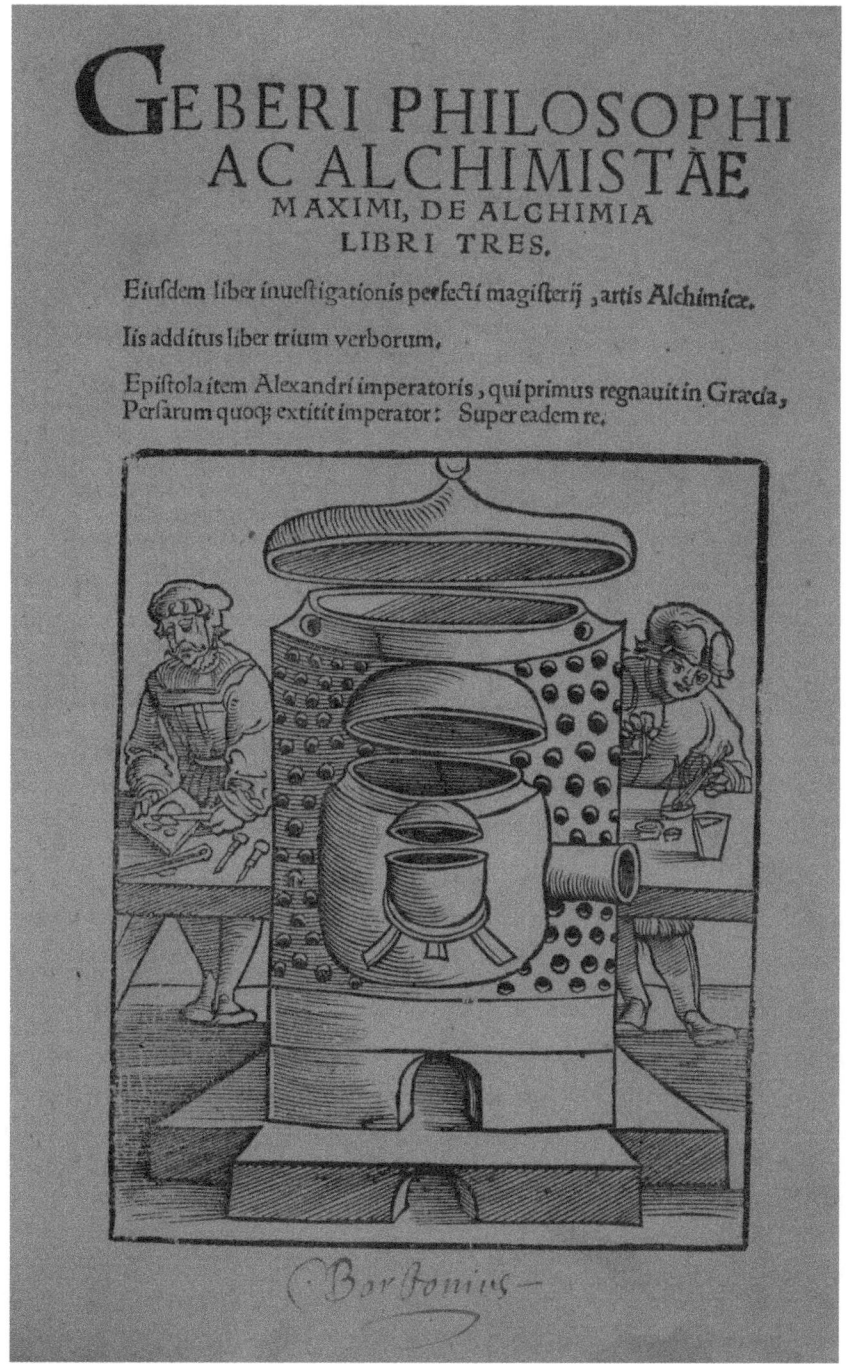

FIGURE 2.1 Title page of an early printed book with works attributed to Geber (Strasbourg, 1531). Credit: Science History Institute, Philadelphia.

FIGURE 2.2 A man at an assaying furnace. From Georg Agricola, *De re metallica* (Basel, 1556). Courtesy Wellcome Collection, CC BY 4.0.

were principally directed toward the transmutation of base metals into gold or silver (Figure 2.2).

A method to separate and purify gold or silver from other metals, cupellation is a process that employs high heat in an oxidizing atmosphere. Metals like lead, copper, and tin are susceptible to oxidation in this environment, whereas gold and silver are not; thus, gold and silver can be refined after the metal oxides are removed by absorption into the body of the shallow porous refining dish known as a cupel. In his text *Theorica et practica*, written at about the same time as the *Summa*, Paul of Taranto described this analytic refining process in detail, noting first the need for a "violent fire" and the addition of a specific amount lead to the metal to be refined in the cupel (Newman and Principe 2002). The author even explained how one can blow on the molten metal through an iron pipe or a reed to remove the impurities fused with the lead.

The major limitation of cupellation was that it could not separate silver from gold or gold from silver, and prior to the widespread use of distillation techniques for the synthesis of strong acids capable of dissolving silver and gold, a different refining process called cementation had to be employed. In short, this technique called for layering thin sheets of a gold and silver alloy with a "cement" that contained common salt and a refractory material like brick dust. The crucible was sealed and then heated, but to below the melting point of the alloy such that the silver would react with the salt, forming what today would be known as silver chloride; this could then be removed to yield purified gold.

These are only two of an array of tests employed within the pseudo-Geberian alchemical corpus, and whereas many other artisans and practitioners simply used these tests to purify metals or to assay the genuineness of a given sample, these texts explicitly employed such tests in a precise manner to understand the fundamental composition of the natural world. As William Newman and Lawrence Principe have shown, the diverse tests that Pseudo-Geber applied were used not only to determine a metal's qualitative characteristics, but also quantitative factors such as specific gravities and whether a metal increased or decreased in weight after undergoing a given process.

The *Summa* exploited such practical processes for distinctly theoretical ends, and in one especially powerful experiment it noted that a given metal could undergo apparently destructive processes like calcination or dissolution in acid, but that the metal could always be reduced back into its original form and recovered. Pseudo-Geber argued that the behavior of metals undergoing such tests could be best understood if they were all composed of very small parts (*minimae partes* or *minima*) of the Aristotelian elements – earth, air, fire, and water – and that these joined together via a very strong composition (*fortissima compositio*) to form the two principles, mercury and sulfur, which composed all metals (Newman 2006a). This experiment would come to be known as a "reduction to the pristine state," and as will become clear in the following section, it had a profound influence on the history of early modern chymistry and matter theory.

Distillation was another ancient technique that had been further developed in the Middle Ages and that strongly influenced early modern chymistry. By the early fourteenth century, alchemists had developed new techniques that allowed the distillation of strong mineral acids capable of dissolving a variety of metals and that proved useful for numerous technologies (Multhauf 1956). The distillation of alcohol was not uncommon in the Late Middle Ages, but it was given a new importance by individuals like John of Rupescissa, who believed that the liquor produced by distilling wine was a powerful medicine capable of vanquishing disease and even prolonging life (Principe 2013). He used multiple names for this medicine, calling it *aqua vitae*, or water of life; the *quintessence*, or fifth essence, comparable to Aristotle's aether or fifth element, which was supposed to exist beyond the terrestrial spheres; and borrowing from earlier alchemical traditions he also called it his elixir. John, however, did not limit himself to medicine derived from wine, and he extended his concept of quintessences to include the distillations of a variety of herbs, animal matter, and minerals – especially those containing mercury, gold, and antimony (DeVun 2009).

This medieval union of medicine with an alchemy that accentuated distillation had an immense influence on the chymistry of the early modern era. The Strasbourg surgeon and apothecary Hieronymus Brunschwig borrowed heavily from John's alchemico-medical thought and brought much of it to the world of print, publishing the first printed manual concerned especially with the distillation of waters for medical uses (Taape 2014). Brunschwig's 1500 *Liber de arte distillandi* (*Book concerning the Art of Distilling*), despite the Latin title, was written in vernacular German and intended for a broad audience (Figure 2.3). It is replete not only with clear descriptions of technical practices and a large variety of alchemical apparatus, but also a multitude of woodcut images that supplemented these descriptions. The new culture of printing allowed alchemical ideas and practices to spread at an unprecedented rate, and numerous other "how-to" books from the time revealed the secrets of nature to an ever-increasing group of interested readers and practitioners. Recipe literature, diverse technical manuals, and so-called "books of secrets" proliferated throughout Europe and were, as William Eamon has convincingly argued, instrumental in the emergence of science (1994). In addition, this proliferation opened doors for women to make important contributions both to alchemy and to chymical medicine (Rankin 2013; Ray 2015) (Figure 2.4).

The individual who most powerfully harnessed and unified these various subjects – medieval alchemy, medicine, craft practice, and print – was the Swiss–German iconoclast Paracelsus, who was born Theophrastus von Hohenheim. In effect, Paracelsus established a system grounded on the practice of analysis that would be taken up by numerous followers, provide a new framework for medicine, and ultimately change the status of European alchemy, elevating it to the level of a philosophy that had cosmological and religious implications.

Paracelsus was a reformer with a knack for controversy and offending authorities. During his itinerant travels throughout Europe, he attacked ancient medical and philosophical orthodoxy, going so far as to publicly burn a copy of Avicenna's *Canon of Medicine*, probably the most respected medical text of the era. Beyond simply lambasting traditional medicine and the Galenism and Aristotelianism taught at universities, however, he proposed to replace these with a system founded on alchemy and a mystical interpretation of Christian religion.

Paracelsus built his new philosophy upon the practice of *Scheidung* (separation), by which he meant the analysis – generally a distillation by fire – of a given natural substance into its constituent parts (Pagel 1982a; Newman 2006a). Based on such distillations, he argued that everything in nature was composed not merely of the ancient dyad of sulfur and mercury, but also of salt, referring to these together as the *tria prima* or three first things. Paracelsus conceived of this analysis of compound bodies not merely as a separation of the constituents of matter, but also as a purification from toxic matter, likewise suggesting that the recombination of separated essences could yield exalted

FIGURE 2.3 Title page from Hieronymus Brunschwig's *Liber de arte distillandi* (Strasbourg, 1512). Othmer Library, Science History Institute, Philadelphia.

FIGURE 2.4 Woman with bellows, from the title page of *Von allen geprenten Wassern* (Nurenburg, 1530). National Library of Medicine.

variants of the original substances. He called this whole process of separation and recombination *spagyria*, and he maintained that the human body had to be understood as a *spagyric* system. This meant that the body was conceived as the microcosm of the natural world at large, and that creation itself was a chymical process, but also that the human body was quite literally a chymical apparatus governed by the separation, recombination, and exaltation of the principles salt, sulfur, and mercury.

For Paracelsus, simple observation of natural objects was insufficient for the acquisition of knowledge in chymistry or medicine, for "to see what is external is for the peasant; to see what is internal, that is to say, what is concealed, is for the physician," and thus, in the following well-known passage, Paracelsus stressed the need to separate the principles from one another through *Scheidung*:

> Thus, if you hold wood in your hand, you have before your eyes only a single body. But this knowledge is of no use to you: the same thing is known and seen [even] by peasants. You must delve and learn until you are certain that your hand holds a sulfur, a mercurius, and a sal. Thus you [must] have the three things visibly, tangibly, and really, each one separated from the other.
> (Weeks 2008: 317–19)

This attitude where the operational or sensory limits of laboratory analysis were deemed the principles or elements of nature has been referred to by historians as the "negative-empirical" principle (Knight 1992; Newman 2006a). Paracelsus was by no means the first to express support for this notion, but his particular imperative to separate natural materials into their constituent principles such that they could be manifestly discerned from one another was especially influential on later chymists. Paracelsus' experimental *spagyria* consequently became so central to the chymical enterprise that, for many practitioners, it defined chymistry itself. Likewise, it became a primary weapon used against competing traditions that the Paracelsians perceived to be unmoored from practice and experience.

In a passage worth quoting at some length, the English Paracelsian Richard Bostocke provided a summary of the foundational importance of the practice of analysis to the Paracelsian worldview, but also an illustrative example of how this could be wielded to attack traditional philosophy and medicine. He wrote:

> The Chymicall Phisition … is ruled by experience, that is to say, by the knowledge of three substanties, whereof eche thing in the great world and man also consisteth, that is to say, by their several Sal, Sulfur and Mercury, y^t by their several properties, vertues and nature, by palpable and visible experience. And when he knoweth the three substanties and all their properties in the great worlde, then after shall he knowe them in man … . The right way to come to this knowleg is to trie all things by the fire: for the fire teacheth the science and arte of Phisicke … . So shall he knowe all things by visible and palpable experience, so that the true proofe and tryal shal appeare to his eyes & touched with his hands. So shall he have ye three Principia, ech of them separated from the other, in such sort, y^t he may see them, & touch them in their efficacie and strength, then shal he have eyes, wherewith the phisition ought to looke and reade with al. Then shal he have that he

may taste and not before. For then shall he know, not by his owne braines, nor by reading, or by reporte, or hearesay of others, but by experience, by dissolution of Nature, and by examyning and search of the causes, beginnings and foundations of the properties and vertues of thinges ...

(Bostocke 1585: Sigs Dv v and r)

The chymists' rhetoric, to put it mildly, was audacious, promising a reorientation of all natural knowledge and medical practice. They trumpeted the superiority of their philosophy of nature primarily because of its reliance upon experience and its unique ability to separate perspicuous principles, and many chymists portrayed traditional philosophers as incompetent and useless sophists who trusted ancient learning before their own senses. Such provocations and polemics were repeated so often and by such a diverse array of chymists in the sixteenth and seventeenth centuries that it is clear these arguments had significant traction and were integral in bringing chymistry to the fore as an archetype of experience in the Scientific Revolution.

It would be wrong to conclude, however, that the Paracelsian promotion of practice and experience was antiphilosophical or anti-intellectual, or even that the followers of Paracelsus were uniformly opposed to traditional philosophy. There were instead a variety of philosophical viewpoints that could accommodate chymical experience, and many individuals reveled in the syncretism that was paradigmatic of the era.

One influential Paracelsian author particularly emblematic of this syncretism was Joseph Du Chesne, sometimes called Quercetanus, physician-in-ordinary to the French King Henry IV. Du Chesne combined Paracelsian chymistry with Aristotelian matter theory and Hippocratic medicine to give the practice of *spagyria* a new philosophical framework, and more than any other Paracelsian, he relied upon the experience of the senses in concert with chymical analysis to make sense of the natural world. As with most other natural philosophers, Du Chesne was a hylomorphist (for which see Chapter 1 of this volume) and maintained that natural materials were explained by the combination of matter and form. He argued that the Paracelsian principles salt, sulfur, and mercury held the active forms of materials, while the Aristotelian elements earth, air, fire, and water provided passive matter, but that chymical analysis could free the active principles from the elements. In this system, the active principles were the direct cause of sensible phenomena throughout nature, from colors to tastes and smells. Du Chesne argued that this understanding of sensible principles was supported by the ancient Hippocratic text *On Ancient Medicine,* which likened the bodily humors to specific tastes, such as salt, bitter, sweet, and astringent. He concluded that the faculties, virtues, and properties of matter are attributed to such tastes, which he referred to as "natural essences" (Du Chesne 1604: 251). These essences were identified with the Paracelsian principles, and because each principle was associated with particular sensible phenomena, a chymical practitioner could carry out a distillation and identify the principles of a given substance. He suggested, for instance, that salts generated tastes, sulfurs produced odors, and mercury spawned colors, and in another schema that sulfurs were known by sweetness, mercuries by acidity, and salts by astringent or bitter properties.

Du Chesne's explanation for the association of the principles with sensible properties relied upon a diverse admixture of Christian theology, Renaissance philosophy, and principally the ideas of Petrus Severinus (Peder Sørensen), author of the influential 1571 *Idea medicinae philosophicae* (*Ideal of Philosophical Medicine*). Du Chesne thus employed astral phenomena, the powers of spirits, and the action of *semina* (seeds) to argue that the three principles only become sensible when the three "formal principles" create a new material "from their own spiritual body [*ex spirituali suo corpora*]," and thereby "animate and adorn it with their properties" (1604: 164). One should not conclude, however, that reliance on this ostensibly esoteric philosophy meant that Du Chesne's practice dealt any less with processes in the laboratory. He noted that these philosophical concepts were "more perceptible in the understanding than by the sense," and proceeded to give extensive descriptions of particular chymical analyses (Du Chesne 1604: 162). He was especially interested in the analysis of salts, and he claimed that one could use fire analysis to differentiate not only the three principles from one another, but even the three different states of salt in a given compound, yielding "a sour, sharp, and mercurial volatile salt," "a sulfurous, nitrous, and sweet salt," and a third "fixed salt" (Du Chesne 1604: 247). Du Chesne used this method of analysis and sensation to describe the chymical makeup of numerous salts, minerals, and metals; he described similar "anatomies" of several plants and spices, and even devoted an entire chapter to demonstrating that the *tria prima* could be found in all living creatures.

Much like Bostocke, Du Chesne found that chymical analysis was a handy cudgel for attacking other traditions. He singled out certain Galenists and Aristotelians, and especially their understanding of matter according to the sensible qualities of hot, cold, wet, and dry, writing:

> The chymists or spagyrists, however, leaving those bare qualities of bodies, sought the foundations of their actions elsewhere, [in] tastes, odours, and colours. At last, by a wise inquisition, they knew there to be three diverse and distinct substances, which are found by a singular artifice in every natural, elemented body: that is, salt, sulfur, and mercury ... For those aforementioned virtual and sensible qualities are to be found in these three hypostatical beginnings, not by imagination, analogy, or conjecture, but in the thing itself [*reipsa*] and the effect. That is to say, tastes primarily in salt, odors in sulfur, and colors out of both, but primarily out of mercury ...
> (Du Chesne 1603: 89)

For Du Chesne, the three principles were analogically related to the three persons of the trinity who were united in a divine "hypostasis," a term meaning something along the lines of an underlying fundamental reality, and thus he declared that salt, sulfur, and mercury were *principia hypostatica* or hypostatical principles foundational to all nature. Helisaeus Röslin had used this Neoplatonic term for the principles several years earlier, but Du Chesne pushed the concept much further and with the most influence, using it, as in the passage above, to upend rival philosophies. While others were busy speculating and engaging in conjectural disputations, the chymists could simply produce "the thing itself" and present foundational, hypostatical reality to the senses. What further argument was needed? Indeed, as the Paracelsian John Webster put it later in the seventeenth century, it was useless arguing with a "Schoolman," for such a traditionalist was so "ignorant in the most admirable and soul-ravishing knowledge of the three great Hypostatical principles ... whose verity is made so evident by Pyrotechny, that he must needs distrust his own senses that will not credit it; but what avails it to sing to a deaf man?" (Webster 1654: 76–7).

FROM *SPAGYRIA* TO *SYNKRISIS*

Du Chesne's attempt to reconcile parts of Aristotle with Paracelsian *spagyria* did not exactly fall on deaf ears; instead, his ideas were swiftly condemned and landed him in an extended and vitriolic controversy with Galenic professors in the medical faculty at Paris (Kahn 2007a). This dispute was an important moment in the history of early modern alchemy, and one individual who entered the fray found a different path for bringing the Aristotelian philosophy into alchemy. This was the polemical German chymist and physician Andreas Libavius (ca. 1555–1616). As Newman has argued, Libavius merged the beginnings of a theory of corpuscular atomism drawn from Aristotle's *Meteorology* with the pseudo-Geberian tradition of alchemy and a new interpretation of Paracelsian *spagyria* (Newman 2006a). Libavius was certainly no friend of Paracelsus, harshly criticizing the latter for his invidious mysticism and heterodox religion; however, he was a serious defender of chymical analysis and synthesis, contending that the modern term *spagyria* was best understood through an etymology that fused the Greek *span*, meaning "to pull apart," with *ageirein*, meaning "to put together" (Newman and Principe 1998). Moreover, Libavius identified these processes of analysis and synthesis with *diakrisis* and *synkrisis*, ancient terms drawn from discussions of Democritean atomism, where *diakrisis* suggested the pulling apart of atoms and *synkrisis* combining them (Newman 2006a).

Our understanding of Libavius' matter theory is partially obscured by the frequency with which he contradicted himself and the bellicose, polemical nature of his writing style (Moran 2007). Thus, knowing exactly what his allegiances to Democritean atomism and corpuscularism were is far from unambiguous. What is nevertheless clear is that his redefinition of *spagyria* and his association of the term with *synkrisis* and *diakrisis* had a major and lasting effect on chymistry throughout the seventeenth century. Libavius' emphasis on these twin practices and his fusion of pseudo-Geberian alchemy with Paracelsian *Scheidung* led to the increasing awareness that substances that had been analyzed could be recombined. This yielded new and powerful insights about gravimetry and matter theory; in particular, it spawned an explicitly atomist experimental chymistry.

One of the most important scions of the Libavian tradition was the Italian-born chymist and physician Angelus Sala, who spent the majority of his career in Germany, where he garnered a certain fame for carrying out several ingenious experiments reliant upon *synkrisis* and *diakrisis*. Most notably, in his 1617 *Anatomia vitrioli* (*Anatomy of Vitriol*), Sala described what was probably the first synthesis confirmed by analysis (Hooykaas 1949: 77–8). He synthesized blue vitriol (i.e. copper sulfate) using copper, water, and spirit of sulfur (i.e. predominantly sulfuric acid), even going so far as to weigh these components. He then decomposed this synthetic vitriol by analysis to recover the original ingredients in the same proportion used in the synthesis. With impressive accuracy, he concluded that water accounts for 33 percent of the weight of vitriol, whereas the modern understanding of the percentage of hydration for copper sulfate is 36.08 percent. For Sala, this demonstrated that vitriol was not a simple or essential unity, but a collocation of particles of copper, water, and acid formed accidentally. Sala argued similarly that the reconstitution of vitriol from its component parts – or, as he put it, its "reduction to its pristine state" – was proof that in the process of analysis and synthesis no transmutation had occurred (Sala 1617: 96–7). It was simply taking a compound body apart and then putting it back together. Elsewhere, Sala described such reductions to the pristine state using the Latin *redintegrare* (Sala 1622: F3r–v). This term and its English cognate, "redintegration," came to be especially important later in the seventeenth century when Robert Boyle popularized an experiment he called the "Redintegration of Niter," but that was actually first discovered by the German chymist Johann Rudolf Glauber (1607–1670; Newman 2006a: 210).

Just as Libavius had argued before him, Sala maintained that the sulfuric acids produced from different sources, including synthetic and natural vitriol, possessed all of the same qualities and were identical (Hooykaas 1983). But extending this line of reasoning and experimentation beyond Libavius, Sala concluded that the individual components of compound substances like vitriol were fixed bodies, explaining that "sulfur always remains sulfur and water always remains water, if their simple substances are regarded without admixtures" (Sala 1617: 79). As Hooykaas has pointed out, this understanding of material entities with definite, constant compositions and distinct properties approaches the modern concept of a "pure substance" (1983: 143), but as we have already seen, Sala's experiment is not out of step with other chymistry of the period. Nevertheless, his anatomy of vitriol is indeed an impressive example of the power of an experiment for shaping theory.

Sala's son-in-law Anton Günther Billich continued this German chymical and corpuscular tradition and went so far as to define chymistry itself in terms of the operations of *synkrisis* and *diakrisis* (Klein 2014b). Billich argued that chymistry should be understood as having an external end directed toward the perfection of medicine through the creation of pharmaceuticals and an internal end defined by *synkrisis* and *diakrisis* that was concerned with resolving compound bodies into their component parts for the purpose of understanding nature and purifying or exalting various substances. Like his father-in-law, Billich believed that these chymical operations demonstrated the truth of atomism. He also drew heavily from the wells of pseudo-Geberian alchemy and Aristotle's *Meteorology*, even titling a chapter in his final book "The object of chymistry is the same as the subject of *Meteorology IV*" (Klein 2014b). Following Libavius, Billich defended a position that fused Aristotle's elements with Paracelsian *spagyria*, wherein the Paracelsian principles were conceived as *principiata*: the first things produced by mixing the Aristotelian elements.

Billich was not the only individual to explicitly define chymistry according to *synkrisis* and *diakrisis*. Most significantly, an influential contemporary in France drew from Libavius to formulate a similar definition, although with important differences from individuals like Billich. In 1612 the Parisian apothecary and chymist Jean Beguin gave several definitions of chymistry in his influential textbook, the *Tyrocinium chymicum*, noting that some call it "Spagyria," by which they refer to its "principal operations," namely *synkrisis* and *diakrisis* (Newman 2006a: 79). Nevertheless, Beguin followed a Paracelsian tradition led by fellow Frenchman Joseph Du Chesne to conclude that the principles were impregnated by celestial seeds and therefore of an ambiguous nature between corporeal and spiritual (Klein 2014b: 357). Libavius himself responded critically to this claim, and Billich extensively attacked and ridiculed Beguin, exposing Paracelsian ideas to the fires of Aristotelian logic. Billich wrote at length, in particular, against Beguin and Du Chesne, defending the corporeality of the principles produced by chymical analysis, arguing for instance that if the principles are not corporeal bodies, "in what manner do they strike the senses, in what manner are they struck?" Billich based this conclusion in part on a quote from the ancient atomist Lucretius, who wrote, "No thing is able to touch or be touched unless it is a body" (Klein 2014b: 358).

DANIEL SENNERT AND THE HOPES OF EXPERIMENTAL CHYMISTRY

The individual who brought this experimental alchemy and corpuscular theory to its early modern zenith, combining Paracelsian *spagyria* with the medieval pseudo-Geberian tradition and pushing *synkrisis* and *diakrisis* to their limits, was the Wittenberg physician and professor of medicine Daniel Sennert. In addition to writing extensively on medicine, Sennert published numerous texts on natural philosophy and chymistry that were widely distributed and translated into other languages. Newman has shown that Sennert's experimental chymistry had a determinative influence on the development of Robert Boyle's corpuscular theory, and other historians have noted the impact of Sennert's atomism, medicine, and philosophy of generation on such luminaries as Gottfried Wilhelm Leibniz and Joachim Jungius (Meinel 1988). Sennert was also one of the first to teach a practical laboratory course on chymistry for university students, just a few years after Johann Hartmann became the first official professor of *chymiatria* at the University of Marburg (Klein 2014a). In addition, Sennert's extensive personal correspondence is an especially rich source for exploring early modern chymical culture and for understanding how chymistry was practiced in the laboratory (Klein 2014a). Unlike most of his medical or chymical colleagues, a large portion of Sennert's correspondence has survived; it only came to light several decades after his death, escaping the editorial intervention of the original correspondents. His letters provide important context for his published works, offering a rare window into a major center of chymistry and chymical medicine in the seventeenth century (Figure 2.5).

By the year 1623, over two decades after Sennert had begun teaching at Wittenberg, he was frustrated with the current state of medicine and natural philosophy, and he wrote candidly about his concerns to his brother-in-law, the Breslau municipal physician Michael Döring (d. 1644). Sennert complained that most physicians and philosophers were complacent because of their overconfidence in tradition, and if admonished for errors, especially by those who relied on experience, they merely react angrily, and "bark just as dogs in the manger." Sennert made clear what group he had in mind for this censure, as well as several of the particular remedies that he believed would restore medicine and natural philosophy. He wrote:

> However, I plainly cannot put aside the care of atoms and the *syncrisis* and *diacrisis* of Democritus; and I entirely believe that injury has been done to that most diligent student of the nature of things by Aristotle, since the Interpreters of Aristotle have ordered us to deduce everything from first qualities, and have mired us in the most dense muck of ignorance, from

FIGURE 2.5 Daniel Sennert. Line engraving by M. Merian (1628). Wellcome Collection, London CC BY.

which we are only barely able to extract our feet. If I were able to withdraw from other duties for at least one month, I would furnish a work in order to exhibit Democritus restored, not without, as I hope, making it fit for posterity.

(Sennert 1676: 590)

Sennert had introduced his commitment to Democritean *synkrisis* and *diakrisis* in his 1619 text *On the Agreement and Disagreement of the Chymists with the Galenists and Aristotelians*, and as the above letter reveals, he believed that his legacy was staked upon these. Sennert hoped that *synkrisis* and *diakrisis* – exemplified by chymical experiments – would institute a groundbreaking reform of medicine, natural philosophy, and chymistry that would forge a *via media* between inflexible traditionalists and impetuous Paracelsians. However, as Newman has shown (2006a), far from abandoning or attacking Aristotle, Sennert followed Libavius' example and grounded his philosophy in the pseudo-Geberian alchemical tradition as well as the reconciliation of Democritus with the *Meteorology*, going so far as to claim that Aristotle was himself a proponent of Democritean *synkrisis* and *diakrisis*. In addition, Sennert's atomism drew extensively from the maverick Aristotelianism of Julius Caesar Scaliger to help direct chymical experiments employing *synkrisis* and *diakrisis* against other theories of matter prevalent among philosophers.

Sennert built his case for atoms on chymical reductions to the pristine state, and while he was not the first to employ these experiments, as we have already seen with Sala, he pushed them further in the service of atomism than had his forebears or peers (Newman 2006a). In one of the experiments that Sennert described, gold and silver were melted together to form an alloy in which the original metals were indistinguishable. The alloy was then placed in *aqua fortis* (mostly nitric acid), whereupon just the silver would dissolve, and after a precipitating agent like salt of tartar (potassium carbonate) was added, the original silver could be recovered by heating in a furnace. Sennert argued that such experiments showed that even apparently uniform alloys were simply a juxtaposition of particles; the silver particles never ceased being silver, either in the alloy or in the acidic solution. Likewise, the metal particles had to be minute, which Sennert demonstrated in some instances of the experiment by filtering the dissolved silver through paper. Already in 1619, Sennert argued that such experiments showed that everything – and not just certain metals – was composed of atoms, and that all Aristotelian generation and corruption was explained by *synkrisis* and *diakrisis*. Over the following two decades, Sennert employed this chymical atomism to explain diverse phenomena in natural philosophy, from the transmission of the soul to the power of fire to produce heat, and also in medicine, including the formation of kidney stones, the transmission of contagious diseases, and the causes of gout (Klein 2014a). In 1636, just one year before his death, Sennert brought his full atomist vision to fruition in his book *Hypomnemata physica*, ostensibly the text that fulfilled his earlier aspiration to "exhibit Democritus restored."

Looking beyond Sennert's published works to his correspondence reveals another important aspect of the Wittenberg professor's commitment to chymical atomism. In brief, his letters with Döring show that the two physicians collaborated for decades in attempting to make and test diverse chymical medicines, whose synthesis and actions they understood through the lens of atomism (Klein 2014a; Klein 2015). They focused their efforts on finding a universal medicine that would have the power to heal all or at least many diseases; in particular, they sought "potable gold" and a universal purgative made from metals and minerals. Potable gold had been discussed by medieval alchemists such as John of Rupescissa, but was extensively popularized by Paracelsus and his followers and usually entailed the dissolution of gold in a noncorrosive liquid. Likewise, purgative metallic remedies containing antimony or mercury were a staple of Paracelsian medicine. Sennert reasoned that potable gold and other "tinctures" that depended upon the resolution of metals into atoms were especially desirable because such drugs could penetrate the body especially deeply. Sennert noted that metallic quicksilver could be swallowed without harm or much effect, and would simply pass through the bowels, but that if it was resolved into particles in a preparation like mercury precipitate (i.e. mercuric chloride), it could insinuate itself deeper into the body and purge more strongly. This view was supported, Sennert argued, by postmortem autopsies of syphilitic patients treated with mercuric remedies, for these revealed deposits of mercury deep within the body, even in the cavities of the bones.

Sennert and Döring are representative of how developments in chymistry and chymical medicine – newly integrated into a university culture – provided a great deal of hope, excitement, and turmoil for learned physicians in the seventeenth century. Many chymical physicians were perpetually seeking out the latest information; they read and exchanged recently published books; and they sought hints of new techniques or recipes that would yield new essences, tinctures, or compounds that could be useful in medicine or natural philosophy. At the same time, Sennert and Döring's pursuit of chymical medicines was driven by especially difficult circumstances, and their experience proved to be fraught. Correspondence shows that the physicians' search for these medicines was motivated not just from philanthropic religious interests or curiosity, but from the fact that both men were desperate for relief from their gout and arthritis (Klein 2014a). Likewise, their communities faced dire social and economic circumstances during the Thirty Years' War. This difficult milieu made chymico-medical information especially valuable, but also led to the proliferation of misinformation and secrecy. Sennert and Döring regularly encountered con men peddling secret remedies, charlatans promising recipes for potable gold or transmutation, and even impostors pretending to be famous alchemists (Klein 2014a). Making matters worse, chymical medicine remained controversial among many physicians, and one of Sennert's contemporaries in Groningen leveled formal charges of blasphemy and heresy related to his atomism and chymical practice. Added to failures in the laboratory or with patients, it is clear that the hopes of early modern chymical physicians were frequently met with disappointment.

VAN HELMONT, THE SOLARY ART, AND SENSUAL PHILOSOPHY

Jan Baptiste van Helmont was another highly influential and controversial chymist who developed a unique view of the Paracelsian principles, recognized the relationship between analysis and synthesis, and used these operations to formulate a corpuscular theory. In addition, he developed and employed new gravimetric techniques that were influential on later chemists, most notably Antoine Lavoisier, whose quantitative experiments on gases were central to the Chemical Revolution (Newman and Principe 2002). Van Helmont had significant reservations about the Paracelsian principles, denying the utility, in particular, of fire analysis. He believed that the principles produced in such analyses were sometimes mere artifacts of the fire, but nevertheless wrote that the terms salt, sulfur, and mercury could still be used, as these were separated by fire from many, but not all, bodies. Van Helmont believed instead that everything was ultimately composed of water, and in place of distillations, he preferred solvents. In particular, he sought what he called the "alkahest," a universal solvent that promised to resolve every tangible body into its prime matter. Even so, van Helmont still lavished praise on the practice of *spagyria*, boasting that Paracelsus' "analysis and synthesis of bodies" was superior to the methods of the schools because it yielded tangible results (Helmont 1652: 482).

Much like Sennert and Sala, van Helmont used cycles of analysis and synthesis to demonstrate that substances that appeared to be uniform were actually compounds formed by the interactions of small particles. Such compounds, he maintained, could be separated into their components, and not only could they be returned to their *pristina initia*, or original ingredients, but also they could be regained in the same original quantity. He demonstrated this by synthesizing glass from salt of tartar (potassium carbonate) and a weighed amount of common sand (Newman and Principe 2006: 77–8). After he had made the glass, he crushed it into a powder and added more salt of tartar, melting these together and then leaving the mixture in a humid atmosphere where it deliquesced, forming so-called "oil of glass" or "waterglass" (i.e. potassium or sodium silicate). Van Helmont then added acidic *chrysulca* (predominantly nitric acid) to the liquid mixture, which yielded a nitrate salt but also precipitated the same amount of sand used to initially produce the glass.

Van Helmont's general interest in weights and measures is well known, and historians often point to his famous willow tree experiment in which he found that, after five years of growth, a tree planted in a pot had gained considerable weight while the weight of the soil had stayed practically the same (Pagel 1982b: 53). This led him to conclude that the weight added to the tree came only from the water that had been added over its lifetime, and that therefore everything is fundamentally reducible to water. Newman and Principe have shown (2006) that much of van Helmont's chymistry was driven by this interest in gravimetry, and combined with his emphasis on analysis and synthesis, this led to his explicit recognition of the concept of mass balance. In short, van Helmont understood, as we still do today, that when substances are transformed in any chemical reaction, the mass at the beginning must be the same as the mass at the end.

In addition to his influential gravimetry, and notwithstanding his critique of fire analysis, van Helmont took cues from Paracelsians like Bostocke and Du Chesne to reinforce chymistry's definition as an analytic art, and then to wield that definition against competing philosophies. In doing this, he helped to give chymistry a unique cultural and disciplinary identity that would endure beyond the seventeenth century. This identity was rooted in chymistry's ability not just to analyze bodies, but also to produce in its analyses concrete principles – the things themselves – that could be sensed. Van Helmont proudly asserted the supremacy of chymistry:

> I praise my bountiful God, who hath called me into the Art of the fire, out of the dregs of other professions. For truly, Chymistry, hath its principles not gotten by discourses, but those which are known by nature, and evident by the fire: and it prepares the understanding to pierce the secrets of nature, and causeth a further searching out in nature, than all other Sciences being put together: and it pierceth even unto the utmost depths of real truth: Because it sends or lets in the Operator unto the first roots of those things, with a pointing out the operations of nature, and powers of Art ...
>
> (1664: 462)

Scholastic ratiocinations and speculations were simply no match for the knowledge revealed by the chymists' analysis, which van Helmont implied had been ordained by God and "given ... unto little ones." Chymical analysis not only opened compound bodies and made hidden things manifest, but also revealed "things themselves [*res ipsas*]" in such a way that they "become social [*socialia*] unto us" (van Helmont 1664: 482).

Van Helmont's followers energetically repeated and even amplified these arguments for chymistry's superiority. In 1651 the English medical reformer Noah Biggs (fl. 1651) wrote that chymistry's principles were "conspicuous by the fire," and that only chymistry could "maketh an investigation into the America of nature, farther then the whole Heptarchy, yea, then the whole Common-Wealth of sciences, all put together" (1651: 57). In the dedicatory letter to *The Art of Distillation*, John French (1616–1657) praised "that Solary Art" of alchemy, whose sun-like brightness outshone all the other arts and sciences. He went so far as to proclaim that only chymistry was "that true naturall Philosophy which most accurately anatomizeth Nature and naturall things, and ocularly demonstrates the principles and operations of them" (French 1651: A3r). Indeed, the chorus of the Helmontians was united in the belief that chymistry was the greatest of the arts or sciences due to its unique ability to separate the real components of natural materials and demonstrate things themselves to the senses. Helmontians employed this emphasis on the practice of analysis, hands-on experience, and sensible principles to denigrate scholastics, Galenists, and anyone with perceived rationalist tendencies. In so doing they elevated their own art and practice to the status of natural philosophy.

By the middle of the seventeenth century, the campaign to define chymistry primarily by its ability to analyze sensible principles had achieved widespread acceptance, and the often implicit negative-empirical understanding of the chymical principles was expressed more and more in definite terms. In his 1660 *Traicté de la chymie*, Nicaise LeFèvre (ca. 1610–1669) declared that unlike the scholastics, who "satisfie by words and meer discourse," the chymist

> will endeavour to bring his demonstrations under your sight, and satisfie also your other senses, by making you to touch, smell and taste the very parts which enter'd in the composition of the body in question, knowing very well that what remains after the resolution of the mixt, according to the rules of Art, was that very substance that constituted it.
>
> (LeFèvre 1662: 9)

LeFèvre explained that chymistry only dealt with questions under the "Judicatory of the senses," and he even dubbed the chymist a *"Philosophe sensal"* (LeFèvre 1662: 19).

Following the Helmontian critique of fire analysis and the increasing prevalence of mechanical explanations of matter, this negative-empirical explanation of the principles was expressed in explicitly instrumental terms. One might not have complete assurance that the principles found after an analysis were the ultimate foundations of nature, but because these were regularly encountered in experiments and were likewise useful in medicine, they were regarded as principles, even if more fundamental principles were one day to be found. This slight skepticism is apparent in Cambridge physician Francis Glisson's *Anatomia hepatis*, where he accepted the expansion of anatomy to comprehend chymical analysis or "Anatomia spagyrica," acknowledging that the chymist's principles could be considered fundamental elements because they were "the last parts man's art or industry can divide things into," and that he preferred "these principles which may be showed to the eye and which really answer in practice," rather than "go further and rely upon mental notions which perchance may prove but mere fancies" (Glisson 1654: 77, 79). The Oxford scholar Thomas Willis took a nearly identical position in his *De fermentatione*, where he explained that he preferred chymistry to either Aristotelian or mechanical explanations because the chymist's "spagyric analysis" resolves bodies into their principles and "determines the sensible parts of bodies" (Willis 1659: 3). To the objection that the *spagyric* principles are reducible to atoms, Willis replied that he would not dispute this opinion if the atomic conceptions could be shown to be real (Willis 1660: 4). While it might be possible to reduce the five principles into more fundamental, uncompounded components, Willis refused to give precedence to other theories until these produced sensible, experimental results on a par with chymistry.

Of course, this understanding of the principles and the hierarchy of the disciplines was not shared by all contemporary chymists; for instance, it was rejected by the famous experimentalist Robert Boyle. Boyle denied that the Aristotelian elements or the Paracelsian principles were the true foundations of nature, and he based his understanding of matter instead on mechanical corpuscles. This corpuscular matter theory was heavily shaped by his reading of chymists like Sennert, but Boyle still argued that chymistry, while a source of useful experiments, was "a Discipline subordinate to Physiques," decidedly deficient in its "Theorical part" (Boyle 1676: 36–7). Boyle recognized the tension that scholars like Willis and Glisson faced, lamenting the fact that mechanists had so far been unable to produce experiments on par with the chymists, and thus a greater number of individuals, "dazl'd as it were by the Experiments of Spagyrists," embraced the chymical understanding of principles instead of mechanism (Boyle 1661: 123).

CONCLUSION: CONTINUITY AND CULTURE

Historians of science long portrayed the rise of the mechanical philosophy as a triumph over competing systems such as chymistry, but far from simply acquiescing or fading away, many chymists continued to fight for epistemic supremacy on questions about the nature of matter. Even throughout the eighteenth century, chymists marshaled the negative-empirical understanding of principles in tandem with sensible demonstrations in the laboratory to provide leverage against mechanists. One group who argued along these lines especially forcefully followed Georg Ernst Stahl, who, in addition to promulgating the phlogiston theory of combustion, had maintained that his chymical principles, having been derived from experiments, were superior to the physicists' principles, which had been invented *a priori* (Fichman 1971; Kim 2003: 171). The so-called French Stahlians capitalized on Stahl's experimentalism to give themselves disciplinary autonomy and a foothold against mechanically inclined peers, and Gabriel Francois Venel's 1753 article on "Chymie" in the famous *Encyclopédie* explicitly followed Stahl and is illustrative of this attitude toward mechanism and the defense of chymistry (Venel 1753; Fichman 1971; Gough 1988).

Venel argued that chymistry (*la Chimie*) was worthy of the attention of philosophers and superior to physics (*la Physique*) because it penetrated "the interior of certain bodies about which Physics knows only the surface and the exterior figure." Whereas physics regarded the qualities of bodies such as colors as mere accidents, the chymist viewed these as physically manipulable substances containing inherent chemical properties. Thus, when the physicist claimed that colors of a plant resulted from mere surface dispositions, the chymist could actually use laboratory analysis to produce a resinous green body responsible for the color. For Venel, such experimental results suggested not that physics was worthless, but that chymistry "simply goes one step further," and that this was especially true with regard to the study of elements. For instance, when the physicists suggested that fire was simply light thrown off a heated body, the chymist could use their analyses and proceed to the level of elements, "that is to say, bodies that [the chemist] can no longer decompose further," and produce the principle of inflammability, or phlogiston, "just as he is able to squeeze water out of a sponge and collect it in another vessel." Much as we have seen before, Venel's argument rested on practices of analysis, the production of things themselves available to the senses, and the negative-empirical principle (Venel 1753).

The prevalence and tenacity of this familiar rallying cry over the course of the early modern period evinces its centrality to the chymists' mentality and cultural identity. Indeed, the chymists discussed in this chapter are marked by significant disagreement regarding the particularities of matter theory, cosmology, and preferred technologies in the laboratory, but the emphasis on analysis unfurled a banner under which many could unite. It is clear that this was an intellectual and ideological movement in which specific beliefs about nature were given prominence, but it was also a significant cultural phenomenon in that it was a collection of distinct practices, perceptions, and values that gave its adherents group autonomy and esteem. Perhaps more than any other discipline in the Scientific Revolution, chymists weaponized practice itself to define and delimit their territory, and within this schema, analysis was the tip of the spear. In effect, adherents defended their analytic discipline as the experimental and practical science *par excellence*.

From medieval alchemy to early modern chymistry and even throughout much of modern chemistry, analysis has been a principal intersection between theory and practice. In this respect, it is evident that the enduring emphasis on analysis was a locus of continuity in the intellectual history of chemistry and was central to the emergence and proliferation of historically significant ideas such as the negative-empirical principle and the concept of mass balance. Beyond this, however, analysis was central to cultural history, and namely the history of the formation of the chemical community, the elevation of the status of chemistry, and the emergence of chemistry as an autonomous discipline.

CHAPTER THREE

Laboratories and Technology: *Chymical Practice and Sensory Experience*

DONNA BILAK

INTRODUCTION

This chapter provides an introduction to the varied cultural landscape of laboratories and technology in England, Europe, and America during the sixteenth and seventeenth centuries. It looks at how a range of practitioners experienced the natural world through "chymistry," an early modern term cognate with – and embracing aspects of – chemistry and alchemy (Newman and Principe 1998). Then, as now, chymistry stands out as the most sensory of the sciences for the ways in which experimental practice yields knowledge through the operator's encounter with color, taste, sound, smell, and touch during a process of chemical change. In situating chymistry within early modern discourse, the term "science" itself merits definition, as it stems from the Latin word *scientia*, meaning knowledge and skill. In the early modern period this implied the synthesis of mind and hand, extending to the chymical laboratory as an epistemic space where intellectual inquiry combined with physical research through the media of texts, materials, and instruments.

Our challenge is to understand what early modern chymists actually experienced in the laboratory. What phenomena did they witness in their chymical vessels? How did they process and interpret what they saw within the context of the cultural and intellectual world that surrounded them? In this chapter, we will explore how early modern practitioners investigated and understood the natural world through the chymical knowledge that they generated in the laboratory. The laboratory space was fundamentally a workroom, a place of chymical production that channeled the use of things (vessels, furnaces, books) together with networks (material, social, geographical) to produce a wide range of chemical medicines (elixirs, salts, oils, balsams). Gold-making, or *chrysopoeia*, was also a medico-alchemical pursuit whose goal was the production of "the great elixir" or "the philosophers' stone." Certain chymists understood these products of metallic transmutation as a panacea: a potent medicament that, by means of a series of complex chemical processes involving the manipulation and purification of matter, could restore health and increase longevity to the human body (Principe 2014c: 24). Chymical practitioners were of numerous types in the early modern period and included, among others, goldsmiths, physicians, apothecaries, princely patrons, and also charlatans (Telle 1994; Kahn 2007a; Kieffer 2014). Attempts to create alchemical gold in the laboratory, and claims concerning metallic transmutation, also brought chymistry under a cloud of suspicion, and some of its practitioners who fell afoul of legal strictures, or their patrons' expectations, stood to lose their liberty or even their life (Moran 1991b; Smith 1994; Nummedal 2019). On balance, the early modern laboratory is defined by chrysopoetic pursuits as well as by the production of medicines destined for use in an individual's medical practice and/or commercial markets (Smith 2006; Klein and Spary 2010). It is this medico-alchemical production that distinguishes the early modern chymical laboratory space from other artisanal workshops that manufactured industrial products such as soaps, perfumes, textile dyes, paint pigments, and glassware (Dupré 2014b; Dupré et al. 2014).

The sixteenth and seventeenth centuries mark an upsurge in experimental science. At this time, drugs made from metals and minerals began to be prepared according to alchemical methods, especially through distillation. The metals and minerals used in preparing chymical medicines were sometimes viewed as living substances, possessed of health-giving vital essences, which the chymical practitioner released through the arts of fire. This notion stems from the medical philosophy of Theophrastus Bombastus von Hohenheim, known as Paracelsus (Debus 1977; Smith 2004: 82–100). Paracelsus privileged the empirical study of nature over book learning, and he sought to supplant the Galenic system of the four humors in the treatment of disease with his own system based on the three principles, or *tria prima*, of mercury (i.e. the quality of moisture), sulfur (heat, fire), and salt (dryness). Galenic pharmaceuticals were preparations of "compounds" and "simples" – medicaments that were designed to possess specific qualities for the treatment of particular diseases, produced by mixing (compounding) materials held to possess a single (simple) quality (García-Ballester 2002). In contrast, the medical philosophy of Paracelsus was rooted in his views about the interrelationship between the universe (macrocosm), the human body (microcosm), and the chemical principles of nature that were held to be embodied in the *tria prima* of mercury, sulfur, and salt. In the production of chymical pharmaceuticals, the practitioner sought to separate the pure and impure parts of a given substance through such operations as distillation, calcination, and sublimation to extract powerful virtues from the vegetable, animal, and mineral kingdoms (Multhauf 1967: 201–36; Moran 2005; Roos 2007). The chymical resolution of matter in the manufacture of medicines produced a wide range of substances in the "spiritualized" forms of liquor (waters, spirits, oils), as well as in "corporified" solid-bodied forms (e.g. salts).

Some well-known figures of the Scientific Revolution were also chymists. The Danish astronomer Tycho Brahe presents a case in point; his castle observatory at Uraniborg not only contained his precision instruments for studying the heavens, its lower floor also held a laboratory equipped with at least sixteen furnaces where he practiced Paracelsian chymistry (Hannaway 1986; Shackelford 1993). A chymist, however, did not have to be a disciple of Paracelsus to practice chymical pharmacy (Moran 1996: 121). Divergent early modern views around the interpretation and practice of alchemy are clearly demonstrated in the example of Andreas Libavius, a seventeenth-century Aristotelian humanist who defended alchemy and was an influential force in establishing the chymical arts as an academic discipline, and who also rigorously debated whether Paracelsians were bearers of new learning in chymistry (Moran 2007). The Flemish iatrochymist Jan Baptiste van Helmont also emerged as an important proponent of a chymical understanding of nature and means of treating disease during the seventeenth century (Pagel 1986; Pagel 2002; van Helmont 1644; van Helmont 1652). Van Helmont is significant to our exploration of early modern laboratories and technology for his promotion of gravimetric quantification in experimental practice as an important analytical tool for tracking reactions during an operation as seen in, for example, the careful weighing of reactants to track gains or losses in specific constituents.

This particular method of chemical analysis in fact played an important role in Robert Boyle's laboratory experiments, which during the 1650s occurred primarily under the tutelage of the Harvard-educated London alchemist, George Starkey. Based on an examination of extant correspondence and laboratory notebooks, we now know the depth of Boyle's reliance on Starkey's chymical expertise, the influence that Helmontian science exerted upon Boyle's experimental practice, and how Boyle presented medico-alchemical processes learned from Starkey as his own (Newman and Principe 2002). An aristocrat and natural philosopher sometimes cited as the "Father of Modern Chemistry," Boyle was an avid chymist dedicated to the pursuit of chrysopoetic alchemy, a fact that necessitates our reassessment of Boyle as an iconic figure of the Scientific Revolution renowned for his mechanical view of nature and the law relating the pressure and volume of a gas (Shapin and Schaffer 1985; Principe 1998; Hunter 2009).

Scholarship in recent years has also drawn back the veil from the alchemical practices of Sir Isaac Newton. As we now know, Newton began chymical experimentation and collecting alchemical books in the late 1660s, and his laboratory work in chymical transformations and *chrysopoeia* was a principal focus of his intellectual activity from the early 1670s until the mid-1690s. Newton maintained a private laboratory in a walled garden located right below his rooms in Trinity College, Cambridge University, where he carried out hundreds of alchemical experiments over the course of three decades; analysis of soil profiles from core samples taken from this site show high concentrations of such substances as copper, arsenic, gold, and mercury (Spargo 2005). For Newton, chymical experimentation constituted a powerful instrument for acquiring scientific knowledge. Furthermore, newly available textual sources, together with laboratory replications of Newton's alchemical experiments, have shed light on the ways in which his chymical investigations related to his work in other intellectual and technical projects (Newman 2018).

Our view of early modern chymistry broadens when we factor in the chymical activities that transpired across the Atlantic in New England. Colonial chymistry presented several intersecting spheres. Iatrochemistry as a healing art was widespread in the Massachusetts Bay Colony, where alchemical knowledge and operations were part of the professional skill set that physicians brought from the Old World. The colonial careers of physicians such as Richard Palgrave in Charlestown, William Avery in Dedham, and John Winthrop Jr in Massachusetts and Connecticut, men who had been educated in England and who brought their knowledge, books, and experience to the colony (Newman 1994: 39–50), offer significant examples. Winthrop, in fact, exerted an important influence on colonial chymistry through his ongoing correspondence with scientific colleagues in Europe and England, with whom he also exchanged materials, equipment, and books. Winthrop saw in alchemy the potential for industrial applications in colonial aspirations to develop mining and metallurgy. His attempts to establish foundries in the iron-rich Massachusetts territories of Braintree, Taunton, and Saugus would, he hoped, finance other schemes, including setting up plantations (Newman 1994: 41; Newman and Principe 2002: 157–9; Woodward 2010). Harvard College was also an active place of experimental chymistry for some of its students and tutors. The seventeenth-century Harvard natural philosophy curriculum explored theoretical considerations of metallic transmutation, and students examined matter theory both in the natural-philosophical curriculum as well as through alchemical discussions in the *Compendium of Physics* attributed to the student and tutor Jonathan Mitchell (Newman 1994: 2–39). We also know about the chymical activities of one student in particular, George Starkey, who, as we have seen, later became Robert Boyle's alchemical tutor. After emigrating to England, he authored treatises on transmutational alchemy under the pseudonym Eirenaeus Philalethes, works that also deeply influenced Newton's chymistry. While a student at Harvard, Starkey was able to develop his chymical interests through correspondence with Winthrop that focused in part on obtaining materials, laboratory apparatus, and literature for his alchemical experiments (Newman 1994: 40). Alongside Starkey, the chymical practices of other Harvard graduates such as John Allin and Gershom Bulkeley have now come to light, and they reveal the significant extent of the college's role in transatlantic social networks and material exchanges in the culture of early modern chymistry (Bilak 2014).

The following sections of this chapter bring into view different aspects of medico-alchemical laboratory processes and products of chymical practice. Our exploration begins with a look at two contrasting representations of alchemical laboratories: an oil painting in the "vanitas" tradition that depicts the orderly laboratory operations of an aging alchemist in his endeavor to produce the philosophers' stone; and a satirical print that presents this alchemical pursuit as a chaotic descent into folly. They serve to situate early modern chymistry as a contested practice that elicited differing social views, yet by the same token both images teem with information about the kinds of instruments and vessels that were used in chymical operations in the laboratory. We then turn to an analysis of chymical technology as described in textual sources, as well as examples of the actual experiences of practicing chymists that are known through extant laboratory notebooks and correspondence. The chapter concludes with a consideration of the laboratory as a symbolic space in which knowledge is concealed through the use of forms as metaphors for medico-alchemical technology and processes.

THE EARLY MODERN LABORATORY IN ART

So, what did the early modern laboratory actually look like? One point of access comes from representations of alchemical practice in contemporary images. These invariably portray the laboratory as a fixed space, although certain textual sources reveal that it could be transportable (LeFèvre 1662: 89, 92–3). It was the presence of vessels used in chymical operations that transformed a room into a laboratory, and these things could be relocated from place to place according to the exigencies of an itinerant practitioner. The laboratory itself could be located in a wide range of places: from princely courts, universities, and monasteries, to simple domestic interiors (Crosland 2005; Smith 2006; Morris 2015).

We can observe the alchemist at work as portrayed in a mid-seventeenth-century oil painting by the Flemish artist, David Teniers the Younger (Figure 3.1). Starting at the upper right-hand corner of this painting, our eye travels along the theatrical sweep of a drawn-back curtain, down to the leather swag hung across a hearth, to the very folds of the alchemist's clothing. Teniers thus guides our sight through a graceful descent into what is in effect a still life composed of retorts, receivers, and crucibles – alchemical vessels made from glass and clay that were staples of early modern laboratory operations. To the right of this collection of chymical equipment we see several books stacked up against a stool placed beside a furnace in which some kind of distillation process is underway. This juxtaposition of texts and instruments represents the close relationship between reading and doing in laboratory practice as the chymist also applied – and tested – knowledge obtained from books in alchemical operations (Principe 1998: 138–80; Timmermann 2008; Gussman

FIGURE 3.1 Interior of a Laboratory with an Alchemist (ca. 1640–1670) by David Teniers the Younger. Oil on canvas. Othmer Library, Science History Institute, Philadelphia.

2015). Teniers has added other interesting visual points to this painting. In the background, we see three figures (likely assistants) engaged in diverse tasks (stirring, grinding, assessing). Hovering above this group is a strange lizard-like taxidermied creature suspended by a rope affixed to the ceiling, a detail relating to the vogue for the collection and display of such wonders of nature – *naturalia* – in early modern museums and cabinets of curiosity (Daston and Park 1998). A small window opens into the laboratory space, and a man leans on its sill watching the activity below him. Teniers' painting also contains an intriguing little interchange involving a mouse nibbling at a candle stub, a little dog who watches it, and a cat in the background who is watching us, a curious *mise-en-scène* that might relate to a vernacular proverb. The alchemist himself dominates the composition of this painting, positioned in between the vessels that actively percolate atop various furnaces to his left and the empty pots and glasses bestrewn along the ground to his right. On the one hand, this positioning of the alchemist eloquently conveys the primacy of the furnace in chymical work. The chymist's sensitive use of the technology of fire, handled through the meticulous manipulation of different degrees of heat that could be generated within the furnace, resolved natural matter into its essence, producing oils, spirits, salts, etc., by means of diverse chymical processes such as distillation, calcination, and sublimation. On the other hand, Teniers' depiction of the alchemist at work in his laboratory alludes to the latter's ongoing, perhaps unsuccessful, attempts to produce the philosophers' stone, signposted by subtle allegorical markers about time: the extinguished candles, the aging alchemist himself, and the hourglass that is set before him, its running sands signifying the passage of time, or of time running out.

We encounter alchemy as the epitome of folly in a mid-sixteenth-century printed image after a drawing by Pieter Bruegel the Elder (Figure 3.2). Here, Bruegel caricatures alchemy as a useless and foolhardy venture. The alchemist, dressed in tattered clothing, sits on a stool before his kitchen hearth, which he has repurposed into a laboratory. Surrounded by assorted flasks, metal clippings, retorts, and other sundry alchemical instruments, Bruegel has rendered the alchemist in the act of placing what is likely his last coin into a crucible for melting while billowing vapors from various receptacles around him catch an updraft in the overhanging hood. The alchemist's back is turned to a scene of domestic chaos, presumably caused by his focus on the furnace as opposed to more pressing family matters: his three children caper about an empty pantry while his morose wife sits in the center of the room, showing the viewer her empty purse. Seated on the floor beside her is a man in a fool's cap (the quintessential symbol of folly), working a pair of bellows over containers whose contents have spilled, and who seems to be singing or shouting as he works. Next to the fool, a scholarly figure sits before a desk with several open books before him. With his right hand, he gestures to the disorderly scene; with

FIGURE 3.2 "The Alchemist" (after 1558) by Pieter Bruegel the Elder and Philip Galle. Engraving; first state of three. The Metropolitan Museum of Art, New York.

his left index finger, he points to words from the book closest to him, "Alghe Mist" – a simultaneous pun on the word "alchemist" and the Dutch phrase for "all is lost" or "all is dung" (Principe and DeWitte 2002: 12). Behind the scholar, a window opens up to show what lies ahead for the alchemist: we see him and his family being received into the poorhouse, the occurrence of which is only a matter of time as per Bruegel's depiction of sand draining through an hourglass placed to the left of the alchemist.

Despite their different portrayals of alchemical pursuits, both Teniers and Bruegel in fact exhibit the same assortment of laboratory equipment: vessels made of glass, ceramic, and metal, as well as crucibles, bellows, tongs, tiles, mortars and pestles, funnels, filters, hammers, sieves, and spoons. This array of things corresponds with enumerations of essential laboratory apparatus found in contemporary books on the subject of medico-alchemical operations, such as LeFèvre's *Compendious Body of Chymistry* and Johann Schroder's *Compleat Chymical Dispensatory* (LeFèvre 1662: 85–9; Schroder 1669: 24–5). Vessels made from glass (alembics, cucurbits, receivers, retorts, vials) were considered the best suited for chymical instruments used in the preparation of medicines because they were inert to most other substances. Moreover, as a transparent material, glassware enabled the chymist to observe what was happening within it. The downside, however, is that glass vessels are prone to break under pressure and heat. Hence a given operation might utilize earthenware pitchers and pots with glazed or tinned inner parts to counter the porosity that is an inherent characteristic of fired clay. Even though such vessels were more durable than glassware, LeFèvre underscores the importance of avoiding sudden temperature changes when working with earthenware pots, as this "would soon cause them to fly in pieces" (LeFèvre 1662: 86–7). Metal vessels (kettles, cauldrons) made from iron, copper, or tin were used only under specific conditions because they could impart foreign qualities to chymical matter that might change its composition and thus the outcome of a process. However, metallic vessels were able to withstand intense heat for long periods, making them well suited for distillations.

Notably, both Teniers' painting and Bruegel's print feature the alchemist working beside a furnace. Wherever the laboratory was situated, whether as a designated or portable space, it was defined by access to a source of heat. The furnace was arguably the most requisite of the chymist's laboratory equipment, as fire was understood to be the chymist's tool for material separations and transformations. LeFèvre aptly explains this when he compares the chymist's use of fire as the agent of separation to the way that an anatomist would use a razor to dissect a body in order to study and analyze its constituent parts (LeFèvre 1662: 19). Indeed, the primacy of fire in early modern laboratory practice is evident in Starkey's stylization of himself as a "Philosopher by Fire" in his *Pyrotechny Asserted*, introduced as "A full and free Discovery of the Medicinal Mysteries studiously concealed by all Artists [i.e. chymists], and onely discoverable by FIRE" (Starkey 1658: title page; Newman 1994).

FURNACE AND FUMES

The French apothecary and chymist Nicaise LeFèvre shows us a wide range of equipment and uses in his chapter "Of the Diversity of all sorts of Furnaces" in his *Compendious Body of Chymistry* (1662: 89–96). This chapter on furnace technology showcases a series of detailed technical illustrations of furnace forms and parts together with the vessels used in specific chymical operations. We are presented with eight annotated furnace-in-action illustrations set in discrete frames spanning four pages, a layout that a modern reader might consider as a kind of graphic novel of chymical operations (Figures 3.3–3.6). Different kinds of furnaces squat or stretch before our eyes. The compact, squarish "Common Furnace for all operations" features adjustable plates for regulating airflow; these are called registers and are marked as "f" in the drawing (see Figure 3.3). Significantly, the dotted lines in this illustration are used to indicate interiority, essentially showing us what is going on inside the

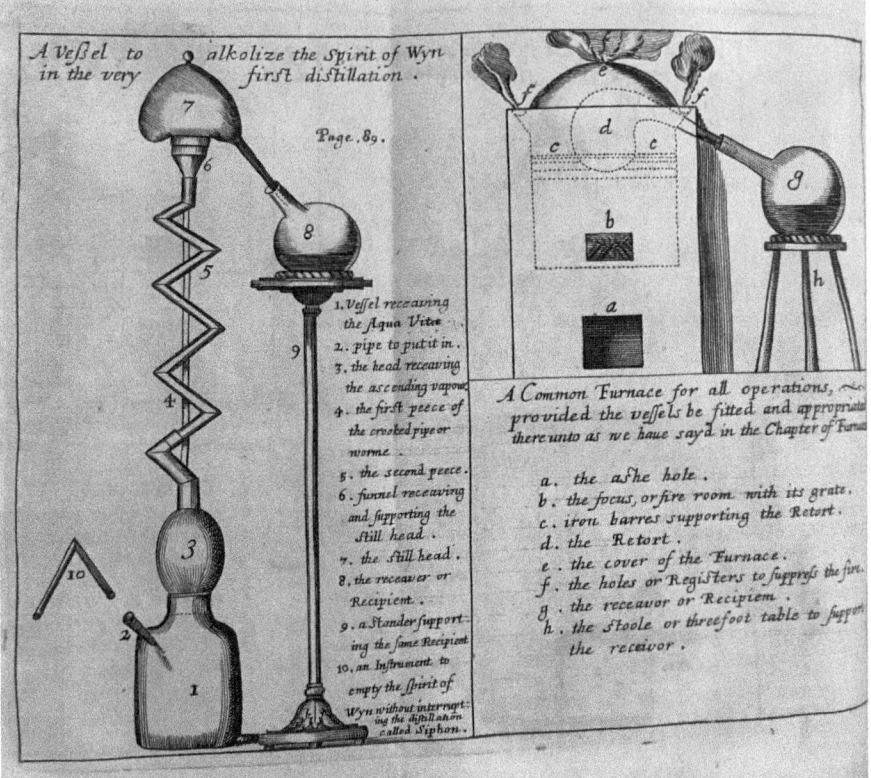

FIGURE 3.3 "Common Furnace" by Nicaise LeFèvre, *Compendious Body of Chymistry* (1662). Othmer Library, Science History Institute, Philadelphia.

furnace. This drawing technique effectively demonstrates the placement of the retort inside this particular furnace (namely, the glass vessel labeled "d"), supported therein by iron bars affixed to the furnace walls ("c"), also showing how the upper body of this retort protrudes from the top of the furnace but remains encapsulated within a heated environment by the furnace's cover ("e"). The "Wind Furnace" presents a very different design (see Figure 3.4). This particular furnace was used for mineral and metallic fusions, vitrifications, and preparations of "regulas" (a term probably derived from *regulus*, referring to the refined metallic component of an ore). The wind furnace features a round turret placed upon a square grate ("f"), elevated from the floor by sturdy square blocks ("e"), so that "the winde and air may have free admission" all around it (LeFèvre 1662: 92). Indeed, LeFèvre's illustration of the wind furnace demonstrates how fire is concentrated all around the matter to be melted, which is placed inside of the crucible that we see sitting in the heart of the furnace ("d").

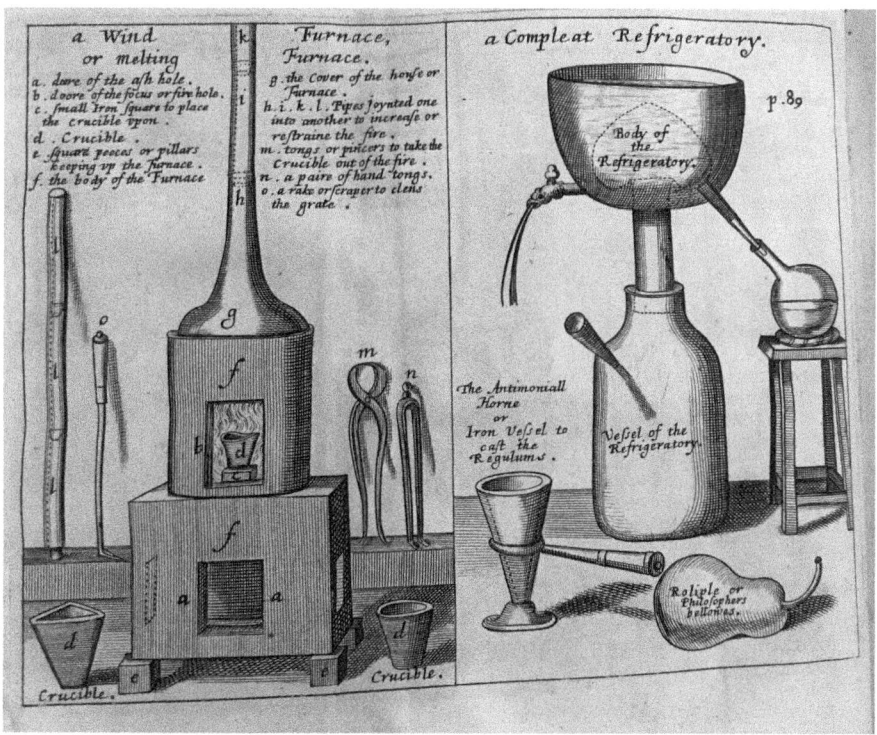

FIGURE 3.4 "Wind Furnace" by Nicaise LeFèvre, *Compendious Body of Chymistry* (1662). Othmer Library, Science History Institute, Philadelphia.

LeFèvre's diagrams give us a sense of how process drives technology apropos of the arrangement of vessels in relation to furnace type. For example, the athanor is a versatile furnace composed of three pieces of equipment: a sand bath, a central tower ("turret"), and a water bath ("bain-marie" or "*balneum maris*"). This tripartite construction facilitates several different operations that could be performed simultaneously (see Figure 3.6). Furthermore, in LeFèvre's various depictions of distillation operations, we observe how the necks of retorts and stills ("Moor's heads") are conjoined with receivers (a.k.a. "recipient" or "*recipiem*") through lengths of crooked or straight pipes (also called "worms"), with the receivers balanced atop stools or small tables. Early modern medico-alchemical distillation adhered to a systematic series of heating and cooling steps to extract essences and purify matter in making drugs in liquid form (Brunschwig 1527; Glauber 1651; Taape 2014). This kind of process is aptly conveyed in LeFèvre's depiction of the "Furnace and Vessels for the Distillation of Waters, Spirits, and Oyles" (see Figure 3.6). Here, we see a

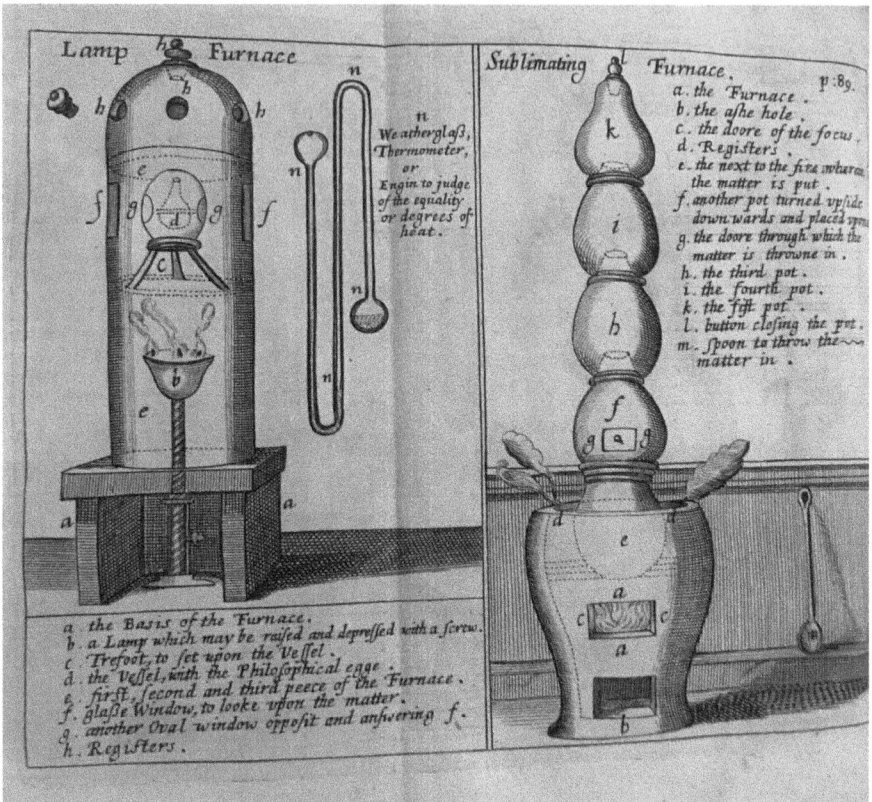

FIGURE 3.5 "Sublimating Furnace" by Nicaise LeFèvre, *Compendious Body of Chymistry* (1662). Othmer Library, Science History Institute, Philadelphia.

large "glasse body for distillation" almost entirely enclosed within the furnace, signified by the dotted lines that outline its body therein ("b"). The top of this vessel protrudes from the furnace, and if we look closely we can see how its mouth fits tightly into another vessel (i.e. the Moor's head marked "c" in this illustration). The Moor's head features a long nose ("g"), which is connected to a pipe (a.k.a. worm) that passes through a barrel containing water to cool and condense the spirits being extracted in the distillation process ("m"); the other end of the pipe/worm connects to a glass receiver balanced upon a block on the ground ("i"), which contains liquid distillate, the product of this process. As a whole, this diagram is a succinct visual representation of LeFèvre's pithily written definition of the distillation process as being "when the matter inclosed in a Vessel [the 'glass body' labeled 'b' in the illustration], drives and sends up vapours in another Vessel [the Moor's head, labelled 'c'], by the help and activity of Fire" (LeFèvre 1662: 77). Indeed, we get a sense of the intensity of heat that governs this particular distillation operation from the small square

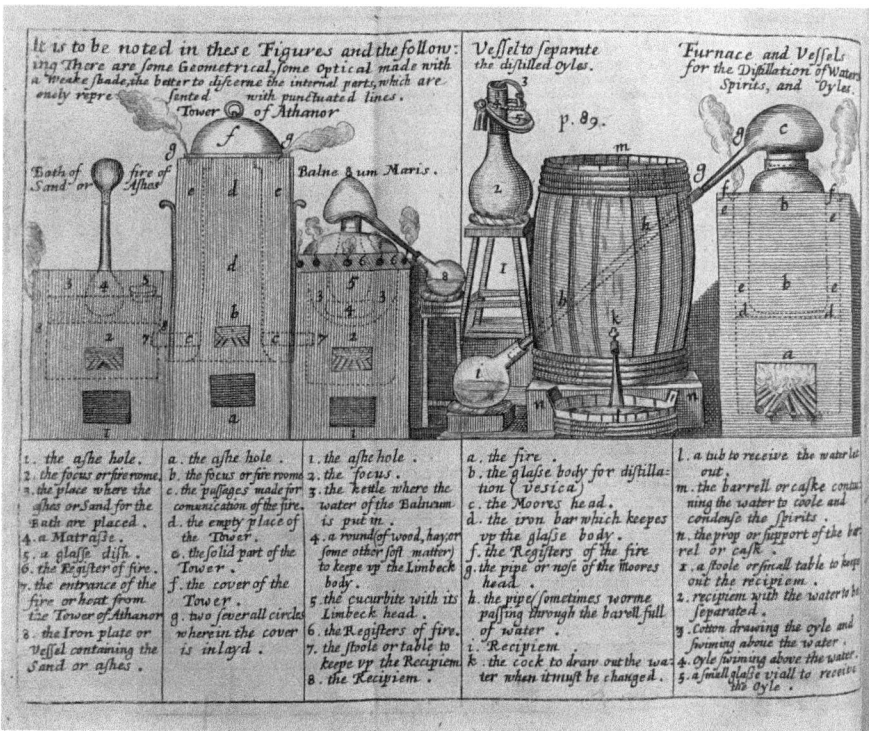

FIGURE 3.6 "Athanor and Distillation Furnace" by Nicaise LeFèvre, *Compendious Body of Chymistry* (1662). Othmer Library, Science History Institute, Philadelphia.

opening at the bottom of the furnace ("a"), which shows faggots of wood and roaring flames heating up the interior of the furnace in which the large glass vessel is held in place by an iron bar ("d"). We can readily imagine how this glass vessel is enveloped by these flames during a chymical process, an action that LeFèvre translates into the jets of steam that we see being released through the registers located at the top of the furnace ("f").

The "Sublimating Furnace" features three vertically stacked and interlocking vessels (labeled "k," "i," and "h" in descending order; see Figure 3.5). These three vessels feed into another vessel ("f"), which acts as a connector between them and the lowermost pot ("e"). LeFèvre renders this bottom pot as three-quarters submerged within the body of the furnace and held in place by iron bars, which are indicated by the horizontal dotted lines beneath it. This connective vessel is further distinguished by a small "door" that is fitted into its body "through which the matter is thrown in" ("g"); this is done using the long-handled spoon that we see leaning against the wainscoting nearby ("m"). A final detail in this diagram of the sublimating furnace: its uppermost vessel is stopped by a "button" to keep vapors from escaping out of it ("l"). In the process of sublimation, a solid substance is changed into a vapor upon being heated, and it forms a solid deposit on the inside of the vessel upon cooling. This diagram of the sublimating furnace shows how fire bathes the lowermost pot that contains the chymical matter, and we are to understand that the fire's action and heat "elevates" the matter once it has transformed into a vaporous state upwards through the stacked vessels. LeFèvre describes the operation of sublimation as resulting in dry "exhalations," of which "at least some part of it ... cleaves and condenses it self into thin and subtile *Flores*" (LeFèvre 1662: 77). These "*Flores*" refer to radiate crystals that resemble "flowers," which can result from the sublimation of certain substances (Newman and Principe 2004: 342). Such sublimates produced from antimony, arsenic, mercury, or sulfur were considered useful additives to certain chymical drugs.

We get a sense of how the chymical technology and substances that LeFèvre describes in the pages of the *Compendious Body of Chymistry* are put into practice with the following example of a chymical experiment undertaken by Gershom Bulkeley at work in his laboratory in Connecticut at the turn of the eighteenth century. Entries from one of Bulkeley's alchemical laboratory notebooks detail a distillation process involving niter, mineral iron, and oil of vitriol that took place in three installments over the course of six days. The overall objective appears to have been the production of *aqua fortis*, literally "strong water" (essentially nitric acid), which was generally prepared by distilling saltpeter with oil of vitriol (sulfuric acid) or with blue vitriol (copper sulfate) itself. That said, Bulkeley appears to have been particularly interested in harvesting the resulting *caput mortuum* (or "dead head") refractory material left in the bottom of a retort or alembic following distillation – basically the dregs (Newman and Principe 2004: 339–40, 345). Interestingly, Bulkeley's experiments in this distillation process revolve around his use of a container that he described as "common potters pitcher" – that is, New England earthenware. This particular item played a significant role in the three successive distillations that took place in 1702 on January 27th and 29th and February 1st.

In the first experimental round, Bulkeley observed that the *caput mortuum* "looked pretty white," and when he performed another operation two days later, the leftover residue was reddish. Bulkeley concludes this second entry with the observation, "Both the Cap. mort [*caput mortuum*] came out easily enough & crumbly but the 2d was not so soft & easy as the first." This seems to indicate mixed results between the January 27th and 29th experiments in Bulkeley's estimation; also, he noted that "the pitcher held very well" in both of these experiments. In Bulkeley's third (and likely final) iteration of this experiment on February 1st, he made some changes in procedure and with the ingredients: he undertook the sublimation of the matter in question, and he added flowers of sulfur. But it turned out that things did not go so well with the third experiment. Bulkeley recorded that the earthenware pitcher in which he had distilled his *aqua fortis* on January 27th and 29th was now cracked "& had leaked a little out into the sand," perhaps indicating that he had an athanor set up in his laboratory in which he performed at least part of this experiment using a sand bath. Moreover, the breakage in Bulkeley's vessel appears to have introduced contaminants, as he goes on to note: "I could not get the Cap. mort cleane off, nor the flowers absolutely cleane: & tis very Probable some might evaporate, the Rec. [receiver] not being luted on." Luting refers to the smearing of a clay-like compound around a vessel to seal it, a preventative measure that would protect chymical matter if the vessel broke from heat.

Bulkeley's specific interest in the *caput mortuum* from this set of experiments, particularly in how he observed changes in color and tactility from one process to the next, indicates that he was basing his experiments on instructions from a printed source. On this point, the works of Johann Schroder present possible candidates; indeed, Bulkeley owned his works (Jodziewicz 1988: 81–2). Schroder's entry for "The Volatile Spirit of Vitriol" in the *Compleat Chymical Dispensatory* describes the *caput mortuum* in terms of preparations (which include calcination, distillation, and sublimation) as well as resultant qualities to look for: "The Salt of Vitriol from this Caput Mortuum, drawn by the Flegm, is the best; other Salts that are fetch'd after three dayes destillation, are only dissolved Vitriol, not deprived of the acide Spirit, you may have a pure Salt, if you Reverberate the Caput Mortuum to whiteness, and Extract it again the common way" (Schroder 1669: 263). By phlegm, Schroder meant a watery substance produced or isolated by a chymical operation (especially distillation), while reverberation referred to a high-domed furnace used for high-temperature heating whose design was believed to drive flames downward (Newman and Principe 2004: 344). Among the diverse therapeutic effects of vitriol, Schroder states that it "heats, dryes, binds, stops vomiting, [and] kills Worms" (Schroder 1669: 261).

Bulkeley's alchemical laboratory notebook is a valuable artifact of his laboratory methods in the production of chymical pharmaceuticals that were likely destined for use in his medical practice. Overall, it is crammed with details about chymical substances, apparatus, and techniques, documenting a range of outputs. Bulkeley made salts, spirits, powders, pills, oils, dissolvents, and elixirs, and in doing so, he worked with different kinds of materials: iron, copper, and antimonial substances, as well as mercury, arsenic, silver, coral, and turpentine. His laboratory notes also specify various chymical processes: calcination, coagulation, sublimation, evaporation, reverberation, and distillation. And Bulkeley's recorded experiments are filled with actions: tasting, weighing, drying, stirring, observing, and, of course, waiting. His notebook describes various retorts (one was silver), cucurbits (he had a glass one and a silver one), a copper vessel, alembics, and receivers. Moreover, in his preoccupation with Helmontian gravimetrics, he made use of different sized scales described as "little" and "great," or "lighter" and "heavy." He also had a little stillyard (steelyard balance) as well as a "greate" stillyard, a wooden spatula, gallipots, glass bottles of different sizes (one is described as square), a blue jug, and an iron kettle, and he regularly used an "oven" (likely a reference to a furnace). His notes record the use of the "naked senses" of sight and touch as instruments of investigation in his experiments (Roberts 2005; Daston and Lunbeck 2011). He even tasted his experimental products, recording the presence, or absence, of various tastes, "lixiviate," "vinegar," "alcalisate," and "urinous." Interestingly, these references to tasting generally occur (when present) at the conclusion of a given entry.

Gershom Bulkeley worked in a designated laboratory space inside of his home in Connecticut. By contrast, his contemporary John Allin pursued an entwined medico-alchemical commercial and chrysopoetic agenda as an itinerant chymist in London who underwent at least four relocations in and around the city from 1664 to 1680 (Bilak 2014). Allin's laboratory moved with him wherever he went, and he always sought accommodations that provided two chambers with one room reserved as a laboratory for a furnace and diverse "chymicall stills" for the "maine worke," a likely reference to the "great elixir." In 1667, Allin took up lodgings in Ratcliff Cross at a pastry cook's house just north of the Thames, and his quarters there featured "a cellar chymney to worke in." Allin relocated to Moorfields in 1670, and a requirement for ready access to a heat source for his chymical work suggests that he read LeFèvre's section in the *Compendious Body of Chymistry* on how to construct a portable furnace "of 1 1/2 foot square" out of bricks and tiles for the chymist on the move (LeFèvre 1662: 92–3). While LeFèvre advised about repurposing vessels (cauldrons with thin sides but "a good strong bottom of Iron plate" for distillation, and "Mettle of a Brest-armour Plate, and forged all of one piece" to serve as a *balneum maris*), Allin owned sundry things such as brass kettles, cauldrons, and presses that could be (or were) repurposed into laboratory apparatus. Allin also possessed a variety of chymical glassware (alembics, retorts, receivers), and such utensils as spoons, spatulas, strainers, sieves, funnels, and tongs. He owned the all-important mortar and pestle for pulverizing ingredients, as well as earthenware crucibles, pitchers, and pots. All told, Allin's setup appears to have included a suite of cushioned chairs, a fire range set, and two little tables (one for each chamber), in addition to which a great chair sat by a table in the area he designated as his "study," where shelves made of large and small deal boards lined the wall(s) for his book collection, which included numerous medico-alchemical works (Bilak 2013). Essentially, Allin's home became an active – and portable – site of science, equipped with the basic furnishings, materials, and apparatus required for chymical work, with his library at hand.

We thus get a sense of the dynamic culture of early modern chymistry though illustrations of laboratory spaces and equipment, as well as through the lived experiences of actual chymists based on archival traces. LeFèvre's various furnace diagrams recall certain items present in the laboratories depicted by Teniers and Bruegel. For example, we see the "Refrigeratory" setup from the *Compendious Body of Chymistry* rendered in action in Teniers' painting, located in the lower left-hand corner near the stack of books (see Figures 3.1 and 3.4). The hand tongs that lean against the wall in LeFèvre's "Wind Furnace" diagram are grasped in the left hand of Bruegel's alchemist, with which he holds his crucible over the fire burning in a small pan beside him (see Figures 3.2 and 3.4). Moreover, we see a continuity of laboratory practices and technologies across the early modern period through the circulation of medico-alchemical texts, as the chymical endeavors of Gershom Bulkeley and John Allin show how authors such as Schroder and LeFèvre may have been used in laboratory operations on both sides of the Atlantic.

The last section of this chapter considers a different kind of medico-alchemical work, one that allegorizes the elemental interactions between mercury, sulfur, and salt (i.e. the *tria prima* of Paracelsian alchemical transmutation, and the substances believed to produce the philosophers' stone). The book in question is Michael Maier's *Atalanta fugiens*, which was specially designed to engage the reader's eyes, ears, and intellect in investigating the chymical secrets of nature. In what follows, we encounter the early modern laboratory not only as a physical space, but also as a symbolic form.

AN ALLEGORICAL LABORATORY

Up to this point, we have explored early modern chymistry as a practical and operative science understood by its various practitioners as the art of material separations and transformations by fire and the chymical manipulation of nature. Yet nature was also perceived as a book in which divine will could be read by those with knowledge and skill. However, this "book of nature" was not accessible to everyone. Knowledge was to be the purview of the worthy, not the vulgar, a contemporary designation derived from the Latin word *vulgus*, meaning "common people" or "rabble." This perspective raises questions about the chymical conventions around concealing information (Principe 2013). In order to protect chymical secrets set down in texts and images from ignorant or unworthy eyes, authors erected barriers around such knowledge using techniques of concealment. This took numerous forms (Newman and Principe 2002: 186–8). For example, alchemical substances or processes were often given "cover names," or *Decknamen*. Other concealment techniques included dispersion (the breaking up and scattering of information across one or many texts), syncope (the abbreviation of an alchemical process and/or omitting information), and parathesis (the needless multiplication of ingredients or processes to confound the vulgar reader). Indeed, the chymists we have considered in previous sections – George Starkey, Robert Boyle, Isaac Newton, Gershom Bulkeley – all took measures to safeguard their written records of laboratory practice through the use of various forms of code (shorthand, symbols, etc.). John Allin's correspondence contains frequent injunctions to secrecy, and certain of his letters requested that the recipient commit chymical secrets to memory before destroying the written evidence.

Michael Maier was a German university-trained physician and alchemist who had been a court *medicus* to the Holy Roman Emperor Rudolf II in Prague between 1609 and 1611. In keeping with contemporary chymical conventions of hiding information, Maier cleverly used his musical alchemical emblem book *Atalanta fugiens* (*Atalanta Fleeing*) to encode alchemical philosophy and laboratory practices. An emblem is a kind of puzzle that is designed to both conceal and reveal information to the reader. It is constructed by the insertion of an image in between two textual elements (i.e. a motto that introduces an enigmatic theme and an epigram that elaborates upon this). The motto, image, and epigram in a given emblem thus work to present a concealed message. In turn, the emblem's message is revealed through the reader's interpretation of its visual and textual interplay, which is aided through recognition of the literary references that the emblem's author has used in its composition. In *Atalanta fugiens*, Maier retells a story from Ovid's *Metamorphoses* about a legendary race between a fleet-footed huntress named Atalanta and her suitor Hippomenes, who won their contest with the help of three golden apples that he received from the goddess of love, Venus. However, in Maier's version, Atalanta and Hippomenes are *Decknamen* for mercury and sulfur, two of the *tria prima*, and in the context of this personification we can also view the golden apple as the third Paracelsian principle, salt (Maier 1618a: author's epigram). Maier allegorizes alchemical philosophy and laboratory arts through a fusion of music, image, and text. *Atalanta fugiens* thus features fifty emblems to engage the reader's "eyes and intellect," paired with fifty musical scores called "fugues" to engage the ears, and each fugue–emblem set is completed by a two-page discourse packed with bibliographic information about alchemical works (Maier 1618a: title page; Nummedal and Bilak 2020). In this way, Maier encodes into his emblems the idea of a metaphorical pathway to wisdom acquired through the chymical investigation of nature's secrets. The reader must decode clues that Maier has embedded in his emblems in order to discover the signs that mark the way forward (Bilak 2020).

In *Atalanta fugiens*, Emblem III can be interpreted as an allegorical representation of calcination, which is the first of seven stages in the chymical production of the philosophers' stone. Maier enumerates these in Discourse

III as starting with calcination, followed by sublimation, solution, distillation, descension, coagulation, and, lastly, fixation (Maier 1618a: 22) (Figure 3.7). The motto in Emblem III reads as an enigmatic command: "Go to a woman washing cloths, you do likewise." Maier develops this theme in Emblem III's epigram, which instructs us to follow the washerwoman's example in cleansing our philosophical matter to remove impurities (Maier 1618a: 21):

> Whoever of you loves to scrutinize hidden doctrines, don't you be idle, bring everything which could be helpful into your model:
> Don't you see how the woman is accustomed to clean stained cloths with hot waters, which she tops off?
> Imitate this woman, and you will not thus be frustrated in your art [i.e. craft, skill], for the wave washes the soil of a black body.

The erudite early modern reader would recognize the literary source underpinning the motto in Emblem III as a reference from the *Tractatus opus mulierum et ludus puerorum* (*On the Subject of Women's Work and Children's Games*), which compares women's work and children's play with chymical processes consisting of mixing, washing, and cooking the four elements (i.e. fire, earth, water, air; De Jong 1969: 68–9). Maier transforms the motto's literary allusion into the depiction of the washerwoman who pours liquid

FIGURE 3.7 Emblem III by Michael Maier, *Atalanta fugiens* (1618). Othmer Library, Science History Institute, Philadelphia.

into a large steaming vat; a spigot at its bottom spews liquid into a smaller vat (interestingly, this resembles LeFèvre's setup for distilling waters, spirits, and oils; see Figure 3.6). Behind the washerwoman we see a vigorously smoking furnace with roaring flames stoked by the bundles of wood piled before it (its low, square form reminds us of LeFèvre's "Common Furnace"; see Figure 3.1). Emblem III plays with the notion of commencing work in its depiction of the first alchemical operation, calcination, whereby matter is cleansed, or "washed," by the licking of the intense flames inside of the reverberatory furnace, which generates the potent fire that reduces matter into a fine calx. Similarly, we see the cloths that are referenced in the motto lying piled up in the lower left-hand area of Emblem III's image, subtly juxtaposed with plies of sandy drifts in the opposite corner, signifying the change from solid to particle that is a characteristic of calcination.

Other emblems in *Atalanta fugiens* also represent different kinds of furnaces. For example, the four interlocking fire-filled orbs in Emblem XVII can be interpreted as an allegorized sublimating furnace, while Emblem XXVIII represents the furnace type known as the *balneum roris*, or the vapor bath. Several images in *Atalanta fugiens*' emblems also depict chymical vessels, such as the mountaintop featured in Emblem VII, which Maier characterizes in Discourse VII as an alembic (Maier 1618a: 34). Discourse XXXIV provides other references to mountains and mountaintops as symbolic representations of cucurbits and alembics, wherein Maier also discusses the receiving flask used in distillation (Maier 1618a: 146–7). Principe interprets the image of the "bedewed house" (*rorida domo*) in Emblem IX as a retort containing a substance known as the "Tree of Diana" (2000). In his laboratory reconstruction of a chemical reaction known to early modern alchemists as "the philosophers' tree," Principe produced the requisite philosophical mercury distilled from various mixtures of metals based on his analysis of recipes in Starkey's laboratory notebooks, demonstrating how the resultant metallic tree that "grew" inside of his long-necked round-bottom flask is a representation of the "philosophers' tree" made from mercury mixed with gold inside of an alchemical retort. While the language of allegory in early modern chymical culture can serve to hide information, it also eloquently describes chymical actions, sensorial perceptions, and the wonder of material transformations.

Maier's use of emblems in *Atalanta fugiens* to both conceal and reveal information relates to a long-standing tradition of engaging the would-be adept in decoding images and/or text to acquire chymical knowledge. A magnificent example of this is the enigmatic and richly illustrated Ripley Scrolls, of which twenty-two are currently known to exist in special collections worldwide (Rampling 2014). Complex allegorical schemes as an alchemical trope began to emerge in Europe by the fifteenth century. Between 1614 and 1622, Maier published several works that made important contributions to seventeenth-century chymical culture. Indeed, Newton found Maier's works to be an excellent resource for alchemical literature, and he considered Maier as one of the best and most useful authors within the alchemical corpus (Figala and Neumann 1990: 34). One example is Maier's 1618 anthology of three principal alchemical texts in the *Tripus aureus* ("The Golden Tripod"), composed of Basil Valentine's "Twelve Keys," Thomas Norton's *Ordinal of Alchemy*, and *The Testament of Cremer*. Maier's *Tripus aureus* provides a bridge between the images in *Atalanta fugiens*' emblems and those in Valentine's "Twelve Keys," a seminal work on transmutational alchemy allegorically enumerated in twelve stages, or "keys," and believed to be by a fifteenth-century Benedictine monk who possessed the secret of transmutation. Maier also produced a cluster of allegorical alchemical works in 1617 and 1618, among which *Atalanta fugiens* stands out as a centerpiece, an elegant narrative expression of mythoalchemy eloquently communicated through music, image, and text (Forshaw 2010). In *Atalanta fugiens*, Maier's "new chymical emblems about the secrets of nature" offer the reader a multimedia experience in the investigation of the secrets of nature by blending music, images, poetry, and science into a paean to wisdom achieved through alchemical knowledge and practice (Maier 1618a: title page).

CONCLUSION

What insights does this look at laboratories and technology offer us about the culture of early modern chymistry? Significantly, what has emerged is that the early modern laboratory was not necessarily a fixed space, that its materials allowed it to be portable, and what happened in it was defined by reading and writing as well as doing. Moreover, new chymical understandings of nature and the human body took shape at the same time that new perceptions of science were being formulated in comparable spaces of experiment, such as the botanical garden, the anatomy theater, and the observatory (Crosland 2005; Ogilvie 2006). Court culture supported the exploration of new technologies, artisanal workshops sought to replicate works of nature, and universities and hospitals began to conduct dissections (Moran 1991b; Smith 2004; Smith 2014). Furthermore, there is the impression of a diversity of practice and practitioners even as we discern a kind of homogeneity underpinning early modern chymistry as it played out on both sides of the Atlantic. Amid these dynamic modes of investigating nature, early modern chymistry reflected a rich, intertextual knowledge system with roots in Greco-Roman and medieval Arabic–Christian intellectual traditions (Ferrario 2009; Ferrario 2010; Martelli 2011), reoriented by Paracelsian and Helmontian ideas that privileged empirical investigation of nature over the written word of classical authority (Clericuzio 1993; Hedesan 2014; Parshall et al. 2015).

This chapter has also shown the ability of text and image – be they instructional, personal, or allegorical – to open up the early modern laboratory as a sensorium of experiences. These engage our sense of sight, yet also extend to our faculties of hearing, smelling, tasting, and touching through the imaginings that these primary sources stimulate with their descriptions of process. While the glinting vessels in Teniers' laboratory draw attention to their different materials (glass, metal, ceramic), this play of light seems to invite us to pick them up and feel their shape and weight. The dotted lines used in Nicaise LeFèvre's diagrams enable us to look inside of the furnace and visualize how its flames would engulf chymical vessels placed therein, eliciting an appreciation for how the technology of fire activates matter and how the design of a furnace or vessel functions within a chymical operation. Michael Maier contrasts the intensity of dry heat and humid heat through the billowing clouds of smoke and steam that dominate the image in Emblem III, evoking the dramatically different sensations of burning and boiling. Gershom Bulkeley's bodily way of knowing was his form of assay, a testing procedure that linked his sensory analysis with chemical analysis in his evaluation of an experiment's progress. In different ways, John Allin, Pieter Bruegel, and Nicaise LeFèvre show improvisation to be another element of early modern chymical culture, in which a kitchen can transform into a laboratory, the cellar chimney in a pastry cook's house can function as a furnace, and discarded body armor can be repurposed into a bain-marie. The experience of early modern chymical culture is not only communicated through printed books and archival remains, it also comes to us through the artist's tools of burin, brush, and pen.

CHAPTER FOUR

Culture and Science: *The Development and Spread of Chemical "Knowledges" across Evolving Cultures and Communities*

ANDREW SPARLING

WONDER, URGENCY, AND THE HOPE OF MASTERY: EUROPEAN CULTURE

Few Europeans in the sixteenth or seventeenth centuries identified themselves as such. People seldom even thought of themselves as having the national identities that we assign to them today – French, German, Spanish, Italian, or English, for example. Modern nation-states were only beginning to take shape, and so too were the national allegiances that we now take for granted. For almost everyone the focus of everyday life remained local: the family, neighbors, village, or town. Anyone outside those local contexts was likely to be judged a foreigner. Nevertheless, in a somewhat flimsy sense, the idea of Europeanness was beginning to emerge (Hale 1994). It rested on two high-flown ideas: that Europeans were the heirs to ancient Greece and Rome; and that the Creator had chosen the Christian community to fulfill a unique destiny.

In the sixteenth and seventeenth centuries, the Ottoman Empire, which was predominantly Muslim, occupied much of Eastern Europe and pressed dangerously at Western Europe's eastern and southern flanks, and certain Western European cities harbored communities of Jews and Muslims. Nevertheless, the dominant culture of Western Europe was Christian. Christendom, though, was anything but unified. From 1517 onward, a massive split, or series of splits, developed in the Western European Church. The resulting conflict, which became known as the Protestant Reformation, resulted, from the mid-sixteenth to the mid-seventeenth centuries, in a succession of bloody civil and international wars. Despite all of this complexity and diversity, however, there remained some Western European notion of a shared Christian cultural identity and therefore of a common fate.

God, many Christians thought, had a plan for history. Although beset by powerful enemies, the Christian community would ultimately prevail against them; and the time for that victory was soon to come. In preparation for that triumph, divine providence would guide the Christian world in learning remarkable new things.

In apparent fulfillment of that cultural destiny, European power was indeed expanding. Only eight years before our period began, Columbus discovered the Americas. By the end of the seventeenth century, European cities, especially in England, Spain, the Netherlands, and France, had established themselves as the hubs of global empires, and far-distant lands paid Europe tribute. As Europe's reach and pull increased, travelers brought home vivid accounts of strange plants, animals, and peoples – as well, sometimes, as the plants, animals, and people themselves. Tellingly, by 1503 the continent on the far side of the Atlantic was already being called "the New World." Horizons were widening, and the frontiers of imagination advanced to keep pace.

At home, Europe's natural philosophers pioneered novel perspectives on the cosmos as a whole and of a human being's place within it. A two-thousand-year-old dogma held that the heavenly bodies all revolved around the Earth and that they were by nature immune to any variation. Two astronomers now shook those assumptions. Copernicus displaced the Earth from the center of things by using mathematics to argue that the Earth was just one among the planets that revolved around the Sun. Galileo turned a telescope to the skies and discerned spots on the disk of the Sun and phases in the disk of Venus. Ocular evidence thereby proved that the heavens were susceptible to imperfection and change.

Historians have pointed to Copernicus and Galileo as pioneers of "the Scientific Revolution," a momentous shift in understanding that involved the wholesale overthrow of time-honored paradigms (Kuhn [1962] 1970). In a literal way, the Copernican revolution pushed human beings and the world they inhabited out of the center of the cosmic picture. All the same, though, it took enormous self-confidence to cast aside a long-held model of the universe and replace it with a new one, based solely on the exercise of human reason, handmade instrumentation, and human eyes. The Earth might have been relegated to the cosmic periphery, but its inhabitants – in particular some of its European inhabitants – considered themselves more than ever in charge. And although the world was a wide and mysterious place, they nonetheless felt up to the task of deciphering, explicating, mapping, cataloging, and colonizing it. They would tame nature, which meant subjecting it to control.

In an intellectual climate where individuals could shatter basic assumptions about the world, as Columbus, Copernicus, and Galileo had done, almost anything seemed possible and within human grasp. The English mathematician and inventor Robert Hooke, for instance, was a genius of apparatus. One could therefore not entirely discount his claim to have invented "thirty several ways of flying" (Aubrey 1898: 410). Curious minds set about collecting accounts of fresh discoveries on the defensible assumption that many of the wonderful stories related about faraway places and things would turn out to be based in fact. Skepticism was not entirely absent: Sir Francis Bacon regarded as a mere braggart the man, "a great dealer in secrets," who asserted that he had thwarted a conspiracy to incinerate England's Queen Mary from a distance with the Sun's rays by means of a giant burning glass perched on a roof (Bacon 1627: 41). On the other hand, the German polymath Samuel Hartlib gave measured credence to the report of a certain Cressy Dymock that "chawed white-bread ... is a kind of Universal Medicine, which Mr. Dymock used with a never-failing success in staunching of blood, 2. swelling of legs, 3. a kind of running pain in the legs, 4. in all manner of bruises, wounds, if the bone be not broken, 5. in biting of a mad dog etc." (Hartlib et al. 2013). Maybe the cure for rabies really did lie in the consumption of specially engineered baked goods. Who knew? All such matters would have to be tested. First, though, they must be collected and set down. It was in the context of this general culture – which approached discovery with urgency and optimism – that various sorts of chemical knowledge developed.

A "SCIENCE" OF CHEMISTRY?

In the sixteenth and seventeenth centuries, no science of chemistry in today's sense yet existed. There was no modern conception of a chemical element, nor of elective affinities, nor of chemical combination. Common laboratory ingredients were not well standardized and canons of experimental evidence were not firmly established. The domains of theoretical and practical knowledge remained loosely knit together, and an institutional framework for carrying out chemical research and communicating chemical discoveries was only beginning to emerge. Chemical investigators often communicated using language that was so obscure that sometimes even fellow experts could not hope to extract a definitive sense. Nonetheless in 1500 Europeans already possessed much chemical knowledge, and by 1700 they possessed a great deal more. More significantly, over the two intervening centuries the social and cultural patterns of knowledge production and dissemination developed in ways that would, in the centuries that followed, make the development of a modern science of chemistry possible.

The word "science," in the sixteenth and seventeenth centuries, simply meant systematic knowledge; and "knowledge," then as now, could be construed in two general ways. Philosophers hold that for something to count as knowledge it must really be true (Feldman 2003). Sociologists argue that when beliefs are systematically arranged and offered within a given cultural context, they too function as knowledge (Berger and Luckmann 1966). Historians of chemistry can profitably apply both perspectives. On the one hand, by our own lights, sixteenth- and seventeenth-century chemical investigators knew much, and we can apply modern scientific understanding to aid us in appreciating not only what they were able to discover, but also how they set about doing it (Principe 1990; Principe 2000; Newman and Principe 2002). On the other hand, the beliefs and practices of the period that are no longer considered scientific or credible today are also relevant to constructing a full historical picture.

If there was as yet no modern science of chemistry, then there was also as yet no "scientific revolution" in chemistry. Historians who have sought to identify a "chemical revolution" have conventionally set the fulcrum of their narrative in the last quarter of the eighteenth century, when Antoine Lavoisier identified and named "oxygen" and discredited "phlogiston" (Conant 1950; Kuhn [1962] 1970: 53–60; McEvoy 2000). In the sixteenth and seventeenth centuries, no similarly compressed or localizable transformation in understanding occurred – no moment that clearly marked a "paradigm shift." Nevertheless chemical knowledge advanced on all fronts, theoretical models became more sophisticated, a partial rapprochement of the practical and theoretical domains occurred, investigators pursued research agendas in the laboratory, and major steps took place toward the institutionalization of chemistry as a discipline. Some chemical discourse became markedly clearer; and thanks to the medium of print, chemical knowledge, together with interest in exploring chemical problems, spread more widely than ever before. While there may have been no scientific *revolution* in sixteenth- and seventeenth-century chemistry, then, there was certainly a scientific boom.

SUBCULTURES

In the absence of a science of chemistry, different brands of chemical knowledge developed in an array of specialist communities. The practical component of sixteenth- and seventeenth-century chemical knowledge fell largely in the domain of artisanal expertise. Dye-makers and dyers, tanners, builders, apothecaries, brewers, vintners, distillers, miners, smelters, smiths, and gunpowderers, as well as fine artists, such as jewelers, painters, and engravers, had developed techniques over generations that embodied chemical know-how, although the techniques did not rest on well-worked-out theoretical rationales. The theoretical component was widely considered a province of what learned contemporaries called *scientia naturalis* or *philosophia naturalis* (natural knowledge or natural philosophy). Natural philosophers endorsed theories of matter, but many of them lacked the practical chemical experience that might have lent their accounts a firmer empirical foundation. In addition to and partly overlapping with both craft know-how and philosophical theory, moreover, there existed a cultural *tertium quid* (third entity), which was a source of both practice and theory. This was the distinctive assemblage of texts and practices that constituted the alchemical tradition. In the sixteenth and seventeenth centuries certain artisans who happened to be alchemists as well as certain philosophers who also happened to be alchemists played a pivotal role in the development of chemical knowledge, partly by bringing into contact with one another disparate subcultures of knowledge producers.

The specialist communities that generated different sorts of chemical knowledge each operated in particular spaces, both literal and virtual. The literal spaces included smelters, smithies, alchemical laboratories, pharmacies, libraries, princely courts, and botanical gardens, such as the Jardin Royal des Plantes in Paris. The virtual spaces took form thanks to written communications: books, manuscripts, and letters. Natural philosophers, for instance, convened meetings to discuss investigations and witness demonstrations, but they also interacted in virtual space. By post or in print, they could communicate discoveries and theories to their peers, disseminate accounts of demonstrations, develop ideas, and conduct disputes (Shapin 1984; Shapin and Schaffer 1985). Constellations of specialist scholars spanned the continent, and however dispersed they might be geographically, those scholars shared a great deal in common with one another in terms of their habits of mind and ways of life. Most important, they shared a common language, namely Latin.

In the mid-seventeenth century, one such scholar, Samuel Hartlib, a German émigré living in England, established a vast European epistolary network of what he termed "intelligencers." Although he was a polymath, who did not hold a university degree, Hartlib nurtured a keen interest in alchemy. He appointed his agents the task of seeking out all manner of scientific and technical knowledge, including chemical knowledge, and every recipe and invention, in whosoever's hands it lay, whether artisans', alchemists', or natural philosophers', and reporting it back to him. One of his intelligencers supplied the recipe, mentioned earlier, for a bread to fend off rabies.

The many groups possessing chemical knowledge interacted with one another in complex ways. A good way to think about the importance of that interplay is with a theory that explains the rise of modern science in the seventeenth century as the result of changing socioeconomic conditions that broke down social barriers between scholars and artisans (Zilsel 2003; Long 2011; Leitão and Sánchez 2017). Encounters always unfolded in specific contexts, where purveyors of different varieties of "local knowledge" (Geertz 1983) could meet and swap techniques and ideas. A site could literally be a specific place, or it could be a figurative one, such as the notional "space" created by an exchange of letters.

The various subcultures that in the sixteenth and seventeenth centuries were concerned with manipulating or understanding chemical substances each possessed their own norms, jargon, and skills. Some (including the natural philosophers and the more educated alchemists) also had theories. In order for productive interactions to take place among representatives of different groups, it was not necessary for the groups to agree about everything. It was sufficient for them to find ways to communicate, in order to exchange ideas, expertise, and artifacts (including books, diagrams, recipes, apparatus, and ingredients). The sites where exchanges took place have been termed "trading zones" (Galison 1995; Galison 1997; Galison 2010). To operate in such zones, members of different subcultures did not all need to see the world in precisely the same terms. They simply needed to improvise a "trading language" – a way of communicating that bridged their cultural differences.

Because chemical theory was less developed and the sum of all knowledge was modest, knowledge-holders had freedom to improvise and innovate. It was also relatively easy to become involved by dabbling, so that princes could become practitioners as well as patrons (Moran 1991c). Furthermore social relations were flexible enough that some practitioners were able to become culturally amphibious: to move from one setting to another, taking on different personae (Rorty 1988). Individuals could even invent novel sorts of personae for themselves and thereby (although sometimes only posthumously) usher a whole new subculture into being.

Two biographical examples help illustrate the potential for the period's mobility and innovation in creating new chemical practices, ideas, identities, and cultures. Johann Kunckel was born in Holstein; his father was a glassmaker, who taught him the trade. Kunckel then trained as a pharmacist, before becoming a court *chymicus* to a duke of Lower Saxony. After 1660 he left that position and established himself as a dealer in medicaments. Then, in the Netherlands, he learned the secrets of Dutch and Venetian glassmaking. Around 1667, now in Dresden, Kunckel took charge of the court alchemical laboratory of Johann Georg II, Elector of Saxony. From Hennig Brand, another merchant–alchemist, Kunckel learned the secret of making yellow phosphorus. In 1667 he fled court intrigue in Dresden and made his way to the University of Wittenberg, where he held lectures for medical students on *Experimentalchemie* (i.e. "experimental chemistry"). He then moved to Potsdam, where Friedrich Wilhelm, Elector of Brandenburg (the "Great Elector"), first hired him to adjudicate the competence of a professed gold-maker. The elector then put Kunckel in charge of the prince's glassworks. There he perfected the manufacture, among other luxury goods, of gold ruby glass. The glassworks held privileges to sell their wares not only to the prince's court but also to Brandenburg's newly founded Guinean Trading Company, which traded glassware, among other things, for slaves, on the west coast of Africa. In 1693 Kunckel entered the service of King Carl XI of Sweden, who ennobled him; and thereafter he was inducted into the Leopoldina, a German learned scientific society, as well as into the French *Académie Royale des Sciences*. He translated, with commentary, an important manual (1612) by an Italian glassmaker, Antonio Neri, and he wrote a number of books of his own on chemical theory and practice.

By any measure, Kunckel was socially, culturally, as well as geographically mobile. In the course of his seventy years or so, he had been an artisan (in two trades); an inventor, technologist, and industrialist; a translator and author; an experimentalist, matter theorist, and learned alchemist; a merchant colonialist speculator; a university lecturer (although he possessed no degree); and a courtier, nobleman, and gentleman-scientist. He spent his entire career moving from one social and cultural zone to another, trading knowledge and ideas as well as more tangible goods.[1] Concretely, we can identify geographically particular "trading zones" in which Kunckel operated, such as his large-scale plant and laboratory on the Pfaueninsel, in the Wannsee near Potsdam. His activities there, though, were but one facet, comparatively easy to localize and depict, of a dazzlingly complex and far-reaching life in chemistry.

A further illustrative example of mobility and innovation relates to the life of Theophrastus von Hohenheim, known as Paracelsus. He lived a much shorter life than Kunckel, and during that life he enjoyed much less worldly success. Nonetheless, no other figure had a greater impact on the development of chemical knowledge in the sixteenth or seventeenth century. Like Kunckel, Paracelsus produced new knowledge by operating across subcultures. In important respects, though, Paracelsus went further. Not only did he move through various milieux, adopting diverse personae according to preexisting models; with conspicuous bravado he also cobbled together novel identities of his own. Although he died penurious and all but unknown, a few followers saw to the publication, decades later, of a huge quantity of his writings. That in turn led to the rise of a novel subculture: the Paracelsians, otherwise known as "chemical physicians" (Clericuzio 2000; Moran 2005; Kahn 2007a).

Paracelsus was a lay preacher, surgeon, and physician (Benzenhöfer 1997; Weeks 1997; Webster 2008). He claimed to hold a doctorate in medicine, but no documentary proof exists that he did, and there is little reason to take him at his word. His father was an illegitimate son of a man from an impoverished branch of the petty German nobility and his mother was a Swiss bondswoman (*Leibeigene*) who was the property of the abbey that controlled the lands where she was born. By law Paracelsus inherited each parent's status: as the son of a bastard he too was considered a bastard; as the son of a chattel he was one too. On his deathbed, aged forty-eight, Paracelsus left a silver cup, his most valuable possession, to the abbot to whom he belonged, as the law required. In life, though, his personal motto had been *Alterius non sit qui suus esse potest* ("Let no one be another's who can be his own man") – a defiant watchword for a person who literally had a legal owner (Figure 4.1).

FIGURE 4.1 Portrait of Paracelsus by the monogramist AH (ca. 1538). Reproduction 1927. Credit: Wellcome Collection, London CC BY.

His father's and mother's legal standing made him doubly ineligible to qualify for a university degree anywhere in the German Empire. He was not even allowed to attend a village school. His first teacher was his father, who also practiced medicine (also, presumably, without university credentials). Several abbots must have recognized in him some promise, because at various times they took him on as a pupil. One was Johannes Trithemius, a humanist, occultist, and alchemist. As a youth Paracelsus then apprenticed in the silver mines of the Tyrol, where he learned metallurgy and alchemy. Even though, as a bondsman, Paracelsus could not in principle travel outside the abbey's territories without the abbot's permission, he spent most of his adult life wandering about Europe (primarily but perhaps not exclusively in the German lands), first as a military field surgeon and later as a physician in search of well-to-do patients to treat. He spent a few disastrous months lecturing to medical students at the University of Basel, before being forced to flee the city.

At Basel he had flouted tradition by lecturing in German instead of Latin. Well before taking up his university post, though, Paracelsus had already earned a reputation for comporting himself in socially transgressive ways. Instead of a physician's red cap, he habitually went bare-headed. Heinrich Bullinger, a Protestant Reformer who had met him, later wrote, "If you had seen him, you would not have said that he was a physician, but rather a carter, and he was marvelously delighted by the companionship of carters" (Erastus [1571]–3: 240). Bullinger was insinuating that the fellow looked rough, and that he enjoyed heavy drinking in low company.

A diminutive man, Paracelsus hauled about with him a huge two-handed sword. Its social message too was double-edged. On the one hand, Paracelsus, despite the appellation "von Hohenheim," was not a knight, so he was not legally entitled to go armed; the sword declared an aristocratic pretension above his actual station. On the other hand, the blade was not a warrior's sword at all. It had instead been the gift of an executioner: a person of the lowest social status, considered unclean (Stuart 1994; Harrington 2013). Despite the fact that Paracelsus claimed the elite status of a physician, he refused to bow to elite norms. In his behavior, dress, and affectation, he in effect declared himself untrammeled by the restrictions that convention sought to impose upon social mobility.

His rebelliousness hardly ended with his self-presentation. Paracelsus also repudiated scholarly medical dogma and natural philosophy. With the exceptions of Hippocrates and the Bible, he rejected all bookish authorities, ancient and modern. In particular he inveighed against the Galenism that was the foundation of university-based medicine and the Aristotelianism that was the foundation of the university curriculum as a whole.

According to the second-century Greek–Roman physician Galen, disease resulted from imbalances in the four bodily humors (blood, phlegm, yellow bile, and black bile). Each of the humors was a composite in which two of the Aristotelian elements predominated (fire, water, earth, and air). In place of Galenism, Paracelsus asserted a novel system of his own, based on three chemical "principles": salt, sulfur, and mercury. The operation of these principles, he claimed, underlay not only physiology but all natural processes. The sulfur and mercury principles were taken over from medieval alchemy. The salt principle, which appears to have been Paracelsus' own idea, was supposed to provide fixity to combinations of the other two. The sulfur principle, which was the principle of all combustibility, would prove, over the next two centuries, to be the theoretical object of a long and many-stranded research tradition investigating combustion. Indeed, beginning almost at the end of our period, in 1697, a fiery principle rather like Paracelsus' would be dubbed "phlogiston" (Stahl 1697).

To counter the authority of traditional, university-based medicine, Paracelsus invoked an alternative source of authority: alchemy. For much and perhaps nearly all of his career Paracelsus was a transmutational alchemist, who professed to know the secret of making lesser metals into gold and who sought out patrons to sponsor him in that endeavor (Kopp 1886; Partington 1961; Sparling 2018). In his medical polemic, meanwhile, he reoperationalized the term "alchemy" and applied it to every manner of material change (Weeks 2008). The human body was an alchemist, because it transmuted food and drink into flesh and bone. A baker was an alchemist, because he changed grain into bread. And a true physician must also be an alchemist, in that he must master and control all the transformations of the body, and to that end be able to prepare every medicine and understand its operation.

In his pursuit of such "alchemical" knowledge, Paracelsus commended in place of what the universities taught the knowledge that belonged to craft-workers and wise women. He urged that genuine physicians (i.e. he himself and anyone who would follow his example) must seek knowledge out in the sites where it could be found. One must travel in order to seek out experts (especially artisans and common folk) and become acquainted by direct experience with the many plants and minerals that God had set in the world for human use. He insisted that in history's final age, which was now upon us, all Christians who worked in the crafts must share their knowledge with one another (and not just with the members of their own trades). He sought out cultural "border zones" and rejoiced in them as sites for knowledge creation. He foresaw no linguistic hurdle in this endeavor. An artisan could communicate what he knew directly – that is, show by doing. To understand what was shown, the learner had only to be pious, unselfish, alert, and receptive. God would provide the requisite insight, in the form of divine illumination. Free trade in technical knowledge would form the cornerstone of Christian community (Paracelsus 1590). Paracelsus may therefore be regarded as the key source of the cosmopolitan impulse that, as

Margaret Jacob has argued, characterized alchemists in the seventeenth century (Jacob 2006; Jacob 2008).

In person Paracelsus fit in poorly among the intellectual crowd. Through his written words, though, he managed, after his death, to strike a chord among a growing number of young physicians from the 1570s onwards. The academic equivalent of pitched battles then ensued, as the old guard sought to repel proponents of the new chemical remedies from establishing beachheads on the faculties of the medical schools. The conflict became particularly heated at Paris (Kahn 2007a). In due course a learned German physician, Daniel Sennert, undertook a judicious and systematic critical assessment of Paracelsianism (Sennert 1619). In doing so he tempered and domesticated chemical medicine and strengthened the Paracelsian version of alchemical matter theory. The "chymistry" of Robert Boyle may be seen as a further major step in the appropriation of alchemy to natural philosophy.

THE CONTINUING TEMPERAMENT OF THE ALCHEMICAL TRADITION

The enduring and peculiar character of the alchemical tradition makes the history of chemistry unlike that of any other science. Although alchemy arrived in medieval Europe at about the same time as Aristotle (both were recovered from Arabic sources beginning in the twelfth century), the two traditions were different in form, content, purpose, and cultural role. Alchemy's origins lay outside Europe. The tradition vaunted secrecy; it cultivated obscurity; it claimed access to a higher, divinely inspired truth; and it never found a place in the traditional university curriculum. For these and other reasons alchemy never fit easily or fully into a natural-philosophical mold. Nevertheless, some natural philosophers became interested in alchemy. They debated its claims and sometimes gave it a place among the arts and sciences. But where some natural philosophers might become alchemists, not all alchemists were natural philosophers.

As other contributors to this volume have pointed out, before the late seventeenth century no clear or consistent terminological distinction existed across the board in the use of the terms *alchymia* and *chymia* or their cognates in the European vernaculars (Newman and Principe 1998). The term *alchymia* and related forms were universal before 1530; thereafter *chymia* and similar variants came into use as scholars, particularly those of a humanistic bent, recognized that the prefix *al-* was merely the Arabic definite article and pruned it as unnecessary. The pioneer of the humanistic usage was Georg Agricola (Rocke 1985). *(Al)chymia* generally referred to one or more of the following three things: an ancient tradition of knowledge and wisdom whose adepts could supposedly produce a "stone" or "elixir" that would transmute lesser metals into gold and/or cure disease and prolong life; the ordinary technical arts of refining metals and synthesizing substances, especially medicines; or all processes of natural or artificial transformation. We may term the last usage, according to which alchemy took place everywhere in the Creation, via natural as well as artificial processes, "alchemy writ large."

Not all authors who wrote about alchemy as the ancient art of producing a stone or elixir were equally interested in metals and medicines. It was even possible to claim that the tradition had never really concerned itself with metals at all. So asserted Gerhard Dorn, a follower of Paracelsus, in his *Clavis totius philosophiæ chymisticæ* (*Key to All Chymistical Philosophy*, 1567). When the alchemists had written about making "gold," Dorn wrote, they had done so "under an equivocation" in order to keep the rabble in the dark about what they were really up to (Dorn 1567: 44). By their philosophical "gold" they had meant various substances, some mineral and some vegetable in origin, whose "temperament" was perfect (i.e. their balance of hot and cold, wet and dry virtues) and whose healing powers were as a result as great as they could possibly be. Such alchemical substances, Dorn claimed, were never compounded but always refined. A genuine medicine must be pure as well as simple.

Alchemy writ large was a sixteenth-century innovation that originated with Paracelsus, for only he and some of his followers applied this broadest meaning to the term. In the 1640s, 1650s, and 1660s, for example, one Paracelsian, Johann Rudolph Glauber, exploited the notion that all material transformations were alchemical in order to dignify winemaking and mundane chemical arts as belonging to fundamentally the same sacred endeavor as the artificial synthesis of gold.

Most often, sixteenth- and seventeenth-century authors meant by *(al)chymia* one or both of the first two senses of the word: the transmutational and the artisanal. Although they did not systematically distinguish the two senses, if they did mean one but not the other the context often made their intention clear. In any case, many if not most writers assumed that the practitioners of the artisanal alchemical crafts were the rustic, less skilled offshoots of the transmutational alchemical tradition. Even those who did not attach much importance to the supposed connection were aware that there ostensibly was one. Therefore even in seemingly mundane discussions of metallurgy, glassmaking, pharmacy, and so on, those arts' reputedly arcane pedigree tended, tacitly or explicitly, to hover in the background.

Here is an example of just how deeply entwined transmutational and artisanal alchemy might be. In 1531 a Frankfurt printer, Christian Egenolff the Elder, decided to take an alchemical text that existed in manuscript, strip the transmutational alchemical material out of it, and publish what remained as a practical handbook of craft knowledge (Kerzenmacher 1531; Eamon 1994: 113–16). Egenolff published the book anonymously under the title *Rechter*

Gebrauch d'Alchimei (*The Proper Use of Alchemy*). The overall tenor of the published book reinforced the impression that the proper use of alchemy was craft-metallurgical. At the end of the book, though, Egenolff appended a moralistic doggerel:

> Eight things follow from alchemy:
> Smoke, ashes, many words, and deceit;
> Deep sighs and heavy labor;
> Wretched poverty and want.
> If you wish to avoid them all,
> Then steer clear of alchemy.
>
> (fol. 37v)

This condemnation was directed against transmutation.

Even though from the text as a whole Egenolff had removed a great deal of material related to transmutation, as well as a preface that the original author or compiler had written, one transmutational recipe remained, which Egenolff attributed to an unidentified adept named Hugo. Even though Egenolff seemed to have tried to make a text entirely about craft-metallurgical alchemy, he did not banish transmutation entirely from its pages.

To complicate matters further, transmutational alchemy was itself always a subject of contestation. Noting the axiom that natural kinds (animal, vegetable, and mineral) could not be changed into one another, some natural philosophers argued that the species of things were fixed (Newman 2004). A keener concern was that, whether or not gold-making was actually possible, it was at least very difficult, and the general run of alchemists were either swindlers or fools. As one late seventeenth-century Austrian wag put it, *Die Kunst ist gut, aber die meisten Künstler sind schlimm* ("The art's fine; it's most of the artists that are terrible"; von Hohberg 1687: 1:77).[2] The image of the alchemical charlatan was a commonplace. The mere existence of polemics and satires deriding alchemical trickery should not be mistaken, though, for evidence that the serious practice of alchemy in any of its senses was in decline (Clajus 1591; Jonson 1612).[3] Indeed some of the most scathing denunciations of the general run of alchemists were penned by people who themselves believed in metallic transmutation and were working to achieve it, but for reasons of self-rectitude, and in order to make a rhetorical point, chose to exclude themselves from their own broad-brush generalizations (e.g. Maier 1617: 14).

Moral concern about alchemists was not of course the exclusive preserve of other alchemists. After the Reformation Catholics were in general more circumspect than were Lutherans. Martin del Rio, though, a Jesuit natural philosopher, simply argued that transmutational alchemy posed no moral hazard so long as only the rich practiced it, because they had no motive to cheat; and should they fail in their enterprise, they would surely not starve (Baldwin 1993: 44).

Del Rio also underscored that a proper alchemist must be humble and God-fearing. Most medieval and early modern alchemical writers would have agreed with him, and they would further have insisted that alchemical success could come only to an alchemist who had those qualities. The pillars on which alchemy stood related both to the object (transmuting metals or the secrets of prolonging life) and to the subject (the alchemist him/herself) performing the work. Theory and practice related to the object; a moral life and divine favor were required of the subject (Mandosio 1990–1; Mandosio 1993).

Crucially, transmutational adepts were imagined to form a secretive community. Thus in the late seventeenth century Robert Boyle alluded to a "sett" (i.e. a group or sect) of "latent [i.e. hidden] Philosophers," "of a much higher order than those that are wont to Write courses of *Chymistry* or other Bookes of that nature." These experts, Boyle was inclined to believe, were indeed "able to transmute baser Metalls into perfect ones, and do some other things, that the generality of *Chymists* confess to be extreamly difficult" (Boyle 1680: Sig. *6r). Boyle meant here by "philosophers" not people who had studied philosophy at university but rather the inheritors of what he and many others believed was a distinctive and privileged corpus of ancient wisdom – a "philosophy" that had been passed on in secret for more than two millennia.

The transmutational alchemical tradition embraced a myth that was quite different from the historical reality. According to the myth, certain pious ancients, before the time of Jesus and even before that of Plato or Aristotle, had possessed a profound knowledge and wisdom, which gave them the power to transmute metals and prolong life. That knowledge had been passed down in secret for more than two thousand years, in part in written form but also orally and manually, by a succession of adepts (some of whom had reputedly lived for centuries). Masters could pass the secret on in person to novitiates, scholars could discover the secret in books, individuals could strive to discover the secret themselves by toiling in the laboratory, and in all cases no answer would come until God chose to provide illumination. There was no room in this narrative for genuine innovation. Alchemical knowledge was elusive, but it was also immutable. The challenge for the would-be alchemist was to find a way back to the knowledge that the ancients had already possessed and that the adepts had faithfully carried on. In reality, of course, the ancients had possessed no philosophers' stone, and no unbroken lineage of masters had passed the secret down to the present day. Instead, the history of alchemy had been syncretic, tangled, fractured, and scattered; swaths of it had simply been obliterated, and what remained was shot through with fictions.

In the sixteenth and seventeenth centuries there really were mysterious individuals who traveled widely, claiming to be the inheritors of ancient alchemical secrets, including in many cases the art of transmuting metals (Principe 1998: 91–137; Nummedal 2002; Nummedal 2007). The majority of such figures probably left no writings behind; these alchemists appear in the historical record, if at all, in accounts written by people who encountered them, or in court cases, when alchemists sometimes stood accused of breach of contract, *lèse majesté*, or simple fraud. We still know relatively little about the motivations, practices, or beliefs of these alchemical practitioners. Some of them were not literate and yet had managed to pick up, by oral transmission and manual demonstration, much of alchemy's cant and many of its techniques. Others were learned men who had read the right books and were prepared to discuss them – perhaps only in exchange for expensive gifts or a series of fees. Meanwhile large numbers of less shadowy practitioners sought (al)chymical knowledge without necessarily laying claim to having already mastered the entirety of the ancient, immutable tradition.

In the course of the sixteenth and seventeenth centuries, practitioners of practical (al)chymy developed and diversified their programs. One important strand of activity was the distillation of alcoholic "quintessences" (*aquae vitae*), in the manner that the fourteenth-century alchemist John of Rupescissa had advocated (DeVun 2009; Matus 2017). Not only monks and apothecaries but also aristocratic women developed recipes for *aquae vitae* and administered their concoctions as medicaments (Rankin 2013). In doing so they enlarged the Western European pharmacopoeia. Other (al)chymists turned away from the volatile products of distillations and instead studied reaction residues, which the Paracelsians in general termed "salts."

On the speculative side, the publication in 1567 of Gerhard Dorn's *Clavis* opened up whole new ways of construing the place of alchemy in the broader field of human endeavor. We have seen that according to Dorn the "chymistical art," insofar as it involved manipulating materials, was solely concerned with producing medicaments, not transmuting metals. That, however, for Dorn, was only half the story. He began by explaining that the "chymistical philosophy" taught in general how to investigate the hidden natures of things. Such a broad notion was very much in the spirit of Paracelsus' wide definition of "alchemy." Dorn went on, though, to develop an account of a whole art parallel to the alchemy of understanding and manipulating materials. It was a program for understanding and refashioning the self of the alchemist (Kahn 2007a: 143–9). This side of the "chymistical art" was "speculative" in the strict sense that it was self-referential (like looking at oneself in a *speculum*, or mirror).

Dorn developed an elaborate schema of correspondences between the seven stages of the laboratory-based alchemical work, which led to medicinal transmutation, and the seven stages of the labor of self-transformation, which led to a sort of spiritual rebirth of the alchemist as a wonder-worker. The seven stages of the speculative side of the "chymistical philosophy" were study, knowledge, desire, perseverance, experience, virtue – and finally power. In the second chapter of the second part of the *Clavis*, the part describing the speculative half of the "chymistical art," Dorn had "Study" exhort the readers, calling on them, in biblical language, to recognize it as "the Way of Truth" and the path to life (cf. John 14:6). People were like stones: they had no eyes to see or ears to hear (cf. Matthew 13:9–16). But people were worse than stones, because people were responsible for their own actions. "Study" called on the members of the "miserable human race" now to take charge: "'May you be transmuted,' it says, 'may you be transmuted from dead stones into living philosophical stones!'" (Dorn 1567: 162). Although Dorn did not explicitly place his novel program for a speculative chymistical philosophy under the rubric of the very word *(al)chymia*, he was in effect opening the way for a new sense of the term.

After Dorn, the notion that alchemy must be properly understood as a discipline that included inward transformation lived on in such writers as Heinrich Khunrath, Johann Arndt, Jakob Boehme, Michael Maier (Tilton 2003), Thomas Vaughan, and Johann Rudolph Glauber. All of these men, with the exception of Boehme and possibly Arndt, enjoyed extensive careers as practicing transmutational alchemists. In 1650, in a publication of his own, Vaughan quoted the passage containing "Study's" peroration for several pages, without naming Dorn as his source (Vaughan 1650: 34–8). A person studied by taking down, poring over, and then incorporating into him/herself others' wise words, which functioned as a delivery system for imbibing the wisdom.

THE PROLIFERATION OF PRINT

Vaughan would have had little trouble finding out what Dorn thought because his words were widely available in print. The advent of the printing press, Elizabeth Eisenstein argued, inaugurated a revolution in European learning by facilitating the transmission of clear information (1979). During the sixteenth and seventeenth centuries at least a couple of thousand imprints relating to chemistry appeared.[4] They varied enormously, though, in how clearly they communicated information, and the most influential and oft-reprinted best sellers, assiduously read by the most technically adept and theoretically learned practitioners of the (al)chymical arts, included some texts that were not clear at all.

Besides clarity, another issue was uniformity. In general, printing tended to standardize texts, whereas the hand reproduction of manuscripts tended to promote variation. Because many alchemical texts had long circulated in manuscript before any were printed, however, there were divergent versions of the same books, and the explosion of print led to many different editions of the same titles but based on different manuscripts. The alchemical texts already involved deliberate obscurities and concealments. When names of authorities or ingredients were inaccurately reproduced, or passages inadvertently deleted, further confusion ensued. Ironically, all of this may have been to the good of the development of chemical knowledge, because it gave practical alchemical enthusiasts of every sort a great deal more to do. Where multiple versions of a text existed, there was nothing for it but to check as many of them as possible and see if one or another made better sense and fit better with whatever results could be obtained in the lab. A virtuous circle often resulted. Discrepancies sold more books, and more books meant more work for people who were trying to learn about (and in some cases write about) alchemy. All of this fed chemical subcultures in contact with one another that became sites of enterprise, competition, and innovation.

Particularly important to the spread of alchemy through print were anthologies. In Latin they included *De alchimia opuscula complura veterum philosophorum* (*Not a Few Short Works of the Ancient Philosophers about Alchemy*), published in 1550 at Frankfurt, in two volumes, by Jacobus Cyriacus; *Auriferæ artis, quam chemiam vocant antiquissimi authores* (*Of the Golden Art, which the Most Ancient Authors Call Chemia*), edited by Guglielmo Gratarolo and published in 1572 in Basel, in two volumes, by Peter Perna; and the monumental *Theatrum chemicum* (*Chemical Theater*), eventually consisting of six volumes, the first four published by Lazarus Zetzner in Strasbourg and the last two by his heirs (vols 1–3 in 1602; vols 1–4 in 1613; vol. 5 in 1622; and vols 1–6 in 1659–1661). The wide dissemination of Gerhard Dorn's ideas owed much to the fact that his writings, including the *Clavis*, were awarded pride of place in the first volume of the *Theatrum chemicum*. There were also impressive collections of material translated into vernacular languages, such as Philipp Morgenstern's translation of the Gratarolo compendium *Turba philosophorum, das ist, Das Buch von der güldenen Kunst neben andern Authoribus* (*The Conventicle of the Philosophers, that is, the Book of the Golden Art, along with Other Authors*), published in two volumes in 1613.

By far the most important collection was Zetzner's *Theatrum*. Its six volumes and nearly five-and-a-half-thousand octavo pages comprised some two million words. It was a masterpiece of seventeenth-century information technology: an entire alchemical library that could fit easily on a writing desk. The title page of each volume of every edition of the *Theatrum chemicum* bore Zetzner's printer's mark, which showed a bust of Athena and the motto *scientia immutabilis* (immutable science). In fact, though, nothing could have been further from the truth than to refer to alchemy either as a single, monolithic science or as immutable. In the sixteenth and seventeenth centuries, thanks now to print, alchemy was diverse and swiftly evolving.

More (al)chymical books were published in the German lands than anywhere else. There were several reasons for this. First, Germany led in book printing overall, because early a thriving community of the requisite artisans proliferated there. Second, literacy was high, in the first instance because German cities were relatively populous. Then during the Reformation Martin Luther as a matter of policy advocated sending all children, girls as well as boys, to school to teach them to read, so that they would be able to read the Bible. Third, mining was a major German industry, especially silver mining. Where there were miners, there were smelters, assayers, and smiths – all chemical crafts. Fourth, the German lands were politically fragmented. There were many princes, and therefore many princely courts. In the age of gunpowder as well as religious strife, wars were getting ever more expensive. To finance their conflicts, the princes were in chronic want of specie. Transmutational alchemists potentially offered princes a potential new source of hard cash (Moran 1991c; Smith 1994; Nummedal 2007). Besides, alchemy offered the tantalizing possibility of being able to transform human hearts and human society (Tilton 2003), which just might usher in a new age of peace (Evans 1973; Moran 1991c). The alchemists discovered that writing and publishing books could be an effective way of seeking out and then acknowledging princely patronage. Not only could a printed book make a handsome gift for a prince, but the book's contents could at the same time help the author wage a campaign of self-promotion.

Up until the early eighteenth century alchemy remained central to chemical discourse, and the most skilled chemical investigators were steeped in its linguistic conventions and habits of thought. Thanks largely to alchemists' efforts, chemistry made substantial gains between 1500 and 1700 both in its technical capabilities and its theoretical conceptualizations. The sixteenth- and seventeenth-century alchemists pursued research agendas, developed fruitful theories, and made empirical discoveries. In doing so they proved effective in bringing chemical practice and theory into closer accord.

FURTHER DIRECTIONS

Probably the most impressive single book on a chemical subject, however, was not alchemical. It was *De re metallica* (*Concerning Metal*) by Georg Agricola (1556), a humanist scholar who became an expert in mining. This fat, handsome folio volume was meticulously written and produced, with detailed engravings. It remains the preeminent sourcebook on the sixteenth-century industries of mining and the production of metals.

In addition to technical manuals like *De re metallica*, another genre of printed books that purveyed chemical information was books of secrets (known in German as *Kunstbüchlein*; Eamon 1994). The *Rechter Gebrauch d'Alchimei* provided one example. Books of secrets typically consisted of snippets of text, as well as some instructional illustrations. (In some cases, the same illustration was used over and over again in the same book, in effect as a piece of clip art.) The snippets included recipes and other short sets of instructions, some gleaned from trades, such as pharmacy or metallurgy; hints on how to repair household wares; and even instructions on how to perform feats of sleight of hand.

Later works relating to mining included *The Restoration of Pluto* (1640) by a Frenchwoman, Martine de Bertereau, the Baroness of Beausoleil. De Bertereau and her husband worked as a couple as mining engineers in the 1620s and 1630s. They found employment under a series of European princes, including two Holy Roman emperors, and traveled widely – even, according to Mme de Bertereau's account, to the great silver mine of Potosí in Peru (today Bolivia). The main thrust of her book was to advocate the exploration and exploitation of France's mineral wealth, but she included a spirited defense of metallic transmutation, which, she insisted, did not require magic or the assistance of demons (de Bertereau 1640: 94–5). She wrote the book while in the employ of France's King Louis XIII and dedicated it to his chief minister, Cardinal Richelieu. "If you authorize what is proposed to you," she promised him in a prefatory sonnet, "… France will soon become a Rich place [*Richelieu*]." Obsequious punning, however, was not enough to put her in the cardinal's good graces. Shortly before his death, suspecting her of sorcery, sedition, or both, he locked both her and her husband up, along with a daughter. Mme de Bertereau died in prison in 1643.

From 1500 to 1700, thanks primarily to the proliferation of printed books and to all the changes that printing brought in its wake, the reach of chemical culture vastly increased. By 1700 far more people and more sorts of people participated in a culture of chemistry than had done so in 1500. Partly this was because a few humanistically oriented scholars had made an effort to make alchemical knowledge more accessible by conveying it more clearly (Moran 2007). At the same time, though, even as textbooks began to appear, alchemical works that made few concessions to transparency continued to proliferate. Whether they were readily comprehensible or not, books that promised to unveil alchemical "secrets" continued to enjoy an ever-widening audience.

Toward the end of the seventeenth century compendia belonging to the genre known as "householder literature" became increasingly ambitious and encyclopedic. Aimed primarily at gentlemen and ladies who managed estates, they were intended to divert as well as inform: they offered household hints but also *Zeitvertreib* (ways to pass the time). These Baroque works were successors to the books of secrets of the previous century. At the very end of our period appeared Andreas Glorez's *Vollständige Hauß- und Land-Bibliothec* (*Complete Library for the Household and Estate*, 1699–1700, in four volumes, with six further installments published as a *Continuation* in the following year). Glorez's compendium contained a great deal of chemical and alchemical information, much in the form of recipes. Among them, somewhat incongruously, was a set of instructions for gold-making. Another recipe proffered a concoction for the treatment of toothache that would supposedly soften the gums in order to expedite an extraction without pain. The procedure involved elements of folk magic: it required capturing a live green lizard, stopping it in a bottle, incinerating the animal in the bottle, and then grinding the body to powder and adding it to the medicine (Glorez 1699: 19). Even more alarmingly, however, the recipe mixed nitric acid with ethanol, which would have produced ethyl nitrate ($CH_3CH_2NO_3$), a powerful and touchy explosive (Priesner 2011: 183–4).

The dawn of the eighteenth century marked the beginning of a new age in which chemistry would develop as a form of public culture. In that century's second half, public demonstrators of chemical phenomena would organize commercially successful spectacles (Golinski 1992). Already by 1700, thanks in part to the publishing success of works like Glorez's, a momentous popularization of chemistry was under way and public cultures of chemistry were taking shape. At this stage people were becoming involved primarily as participants rather than spectators. Well-to-do laypeople read about (al)chymy in their armchairs, conversed about it in their gardens and parlors, and dabbled in it in their outbuildings or kitchens.

(Al)chymical ideas and practices circulated among more people than ever before. That meant not just the diffusion of existing "knowledges" and the creation of fertile conditions for generating new ones, but also, quite broadly, the emergence of fresh patterns of human experience. To adjust at home, for instance, the procedure for making a new, improved kind of laundry soap, by tinkering with a recipe taken from a book, was to live the entire undertaking – heating, mixing, and stirring; trial and error; the hope of cleaner clothes and the risk of chemical burns; the weariness of labor and the triumph of success – in a new way: as an inquisitive, resourceful, and well-read householder, engaged, with other members of a nascent public, in an enterprise of improvement (Hacking 1986; Csordas 1993). From the sums of such sets of lived experiences, many of them quite everyday and ordinary ones, the subcultures of chemistry grew, evolved, and spread.

As they did so, the myth would endure that by understanding and controlling nature human beings were fulfilling a providential role. Henceforward the pervasiveness, effectiveness, and sheer complexity of chemistry, together with the other sciences, would tend to eclipse to some extent the religious tenet that historically had driven faith in scientific advancement. Nonetheless, the belief that scientific progress was inevitable would long remain an article of faith; it would help spur investigators along to genuine discoveries.

CHAPTER FIVE

Society and Environment: *The Social Landscape of Early Modern Chemistry*

WILLIAM EAMON

The history of early modern chemistry is usually defined by larger-than-life figures such as Paracelsus and Robert Boyle and by the battles among rival opinions that led ultimately to what Allen Debus referred to as the Chemical Philosophy, then onward to the final victory of the mechanical philosophy of nature (1977). But Paracelsus was hardly a typical early modern chemist (Moran 2019). Most practicing chemists probably had never heard of the man. Practicing chemists in general cared not a whit about chemical philosophy or the battles raging among competing alchemical schools on the nature of matter. Most early modern chemists didn't even call themselves chemists, or alchemists. Instead, they called themselves apothecaries, dyers, metallurgists, potters, and a host of other artisanal names, because that's what they were.

Early modern chemical practitioners were not for the most part alchemists seeking quasi-mystical transmutations or the mythical philosophers' stone. Instead, they sought – and accomplished, often with consummate skill – transmutations of matter that yielded practical objects and drugs and sometimes magnificent works of art, like the refined tin-glazed maiolica pottery produced in Italy that enchanted early modern collectors. Few of them wrote erudite works in Latin revealing the mysteries of alchemy. In fact, few could read or write Latin. Chemistry was practiced in workshops, studios, kitchens, and barnyards as well as in courts, academies, and laboratories.

Early modern chemists typically didn't teach at universities because chemistry wasn't a part of the university curriculum until the seventeenth century (Moran 1991a). The first chair of pharmaceutical chemistry was founded at the University of Valencia in 1590s, but it was occupied for only a year (López Piñero 1977; López Terrada 2005). Other university chairs of chemistry were founded somewhat later, at Marburg, Paris, Leiden, and elsewhere, but not until the second half of the seventeenth century. As Bruce Moran notes, it took a "cultural clearing" to make way for chemistry as a didactic discipline before it could assume a place among traditional disciplines such as medicine. Such clearings are rarely tidy, and they often fuel hostile polemics and violent acts of character assassination (Moran 2007).

Of course, alchemists also did chemistry; but those who identified themselves as alchemists were by far the minority among early modern chemical practitioners, and they often disagreed over what an alchemist was. Was an alchemist one of those wretched, deluded fanatics in tattered clothing sweating over impossible dreams such as discovering the elixir of life or the philosophers' stone, as portrayed in some genre paintings of the day? Or was an alchemist a practical man who exercised "the right use of alchemy," as a sixteenth-century craft manual instructed, such as making good steel? Of course, it's misleading to use the modern term "chemist" to describe early modern chemical experts, or "chemistry" for the myriad arts they practiced. But the term "alchemist" can be just as misleading. Historians generally agree that the anachronistic word "chymist" better describes the diversity and amplitude of the activity contemporaries called "chymistry" (see Chapter 1). When we think of early modern chemistry, or chymistry, perhaps we ought to heed the advice that the sixteenth-century female alchemist Isabela Cortese gave aspiring adepts. Lady Isabela was an avid alchemical practitioner who had traveled all over Italy and Eastern Europe in search of new techniques and alchemical secrets. Yet while she was an experienced adept, she was contemptuous of alchemical theory. If you want to practice the art of alchemy, she advised neophytes, "Don't follow the teachings of Geber or Ramon Lull or Arnald of Villanova or any of the other philosophers, because their books are full of lies and riddles" (Cortese 1561). As Moran observes, "Alchemy was never altogether anything that people *believed* in; it was something that people *did*" (Moran 2005: 10, emphasis in original).

THE PRACTICE OF CHEMISTRY

Who were the early modern chymists? By and large the people who worked daily with chemical processes and tried to understand and improve them were interested in practical results, not theoretical explanations or speculation about the nature of matter. They included craftsmen of myriad backgrounds and trades. Chemical processes were prominent in a wide variety of crafts.

And though historical research on early modern chymistry has focused (quite naturally) on those who contributed to the flood of early modern alchemical literature, most practicing chemists published nothing. The practice of chymistry was about producing things for use and leisure, whether a drug, a steel tool, or a bottle of grappa. Larry Principe writes, "Chymical discourse and practice took place in academic chambers, in courtly settings, in private homes, in commercial establishments, and in noisy workshops" (2014b: 158). And one might add, chymistry also took place in barnyards where saltpeter-men collected weathered manure to refine into mysterious saltpeter; in kitchens, where housewives distilled elixirs in simple stills; and in apothecary shops by pharmacists who transformed traditional herbal drugs into exotic potions.

The development of chemical industries in the sixteenth century was driven by rapid economic growth, an expanding market economy, and a booming demand for luxury goods. Following the demographic catastrophe of the Black Death, Europe's population began to recover after 1460, and it increased sharply in the sixteenth century. The expansion of consumption resulted in an increase in the volume of manufactured goods produced and traded, the development of new industries, and the growing amount of goods imported from the non-European world (Musgrave 1999: 61). As manufacturing moved out of the household and into places of concentrated manufacture such as workshops and factories, workers followed. Agriculture no longer afforded people with a secure living, and early modern industrial expansion provided welcome economic opportunities for agrarian workers forced off the land. Most of the demographic growth took place in cities, where economic opportunities provided by industry drew labor from the countryside. Chemical enterprises were thus vehicles of social mobility. Learning new chemical technologies such as distillation could transform a person from a landless nobody into a respected citizen of an urban community.

The early modern capitalist economy was also an increasingly global economy. Many Europeans who traveled to America went seeking economic opportunities and found them in chemical industries such as metallurgy. Traditional trades such as dyeing were transformed by the introduction of new dyestuffs such as cochineal. But of all chemical processes, distillation reigned supreme because it was at once a practical science, a philosophical science, and a noble science.

A GOLDEN AGE OF DISTILLATION

The sixteenth century was a golden age of distillation, not just as measured by the proliferation of treatises on the art, but also by its ubiquitous physical presence. Distillation was employed in the pharmacies because it promised to produce wonder drugs by separating out the essential "active" ingredients from the dross left behind in the cucurbit. Distillation was also prominent in metallurgical industries, where it was used to make parting acids for dissolving precious metals, to extract sulfur, and to manufacture sublimates. On the streets and in the shops and marketplaces of cities and towns across Europe, people could buy essential oils, brandy, perfumes, and medicines made by distillation. They could also see perfume made by distillation. Giovanventura Rosetti described and illustrated the sundry distilling alembics that perfumers employed in their shops (1973: 227). Even charlatans sold distilled essences in the market squares. The alembic was almost ubiquitous in the cities and towns of early modern Europe. If you looked, distillation was all around you.

As with most technologies, distillation traveled well. Many of the distillation vessels used in Venice and Naples were also used in Germany and elsewhere. The familiar shapes of alembics and cucurbits soon became standardized. The spread of the technology of distillation was facilitated by the printing press. Treatises on distillation proliferated, spreading the art. European distillation also spread globally with the expansion of the Spanish Empire in America and Asia. Among other uses, distillation was employed in the process of amalgamation to extract silver, helping make the mines of Potosí the main supplier of European silver (Figure 5.1).

While the visual experience of seeing distillation in a workshop became part of everyday life, the idea of distillation served as a powerful metaphor of the separation of the pure from the impure in the wider social and religious sphere. Laurence Andrew, the English translator of Hieronymus Brunschwig's famous treatise on distillation (1973), defined the art in a way that rings of a sermon:

> Distilling is no other thing, but only a purifying of the gross from the subtle, and the subtle from the gross, each separately from the other, to the intent that the corruptible shall be made incorruptible, and to make the material immaterial, & the quick spirit to be made more quicker, because it should the sooner pierce & pass through by the virtue of his great goodness and strength that there in is sunk and hid for the conceiving of his healthful operation in the body of man.

In an age of intense religious conflict, this language resonated. The separation of the saved from the damned at the end of time was a theme emphasized by preachers everywhere, whether Catholic or Protestant.

Distillation was also a frequent topic of discussion in philosophical circles. It was a new chemical technology; the ancients were thought to be ignorant of the art, and for that reason it was inherently of interest to humanists and scholars. Novelty always provokes curiosity. Moreover, distillation was one of several

FIGURE 5.1 Pharmacy Transformed by Chymistry. Traditionally the apothecary practiced the simplest form of chemistry, grinding and mixing herbs and minerals to fill physicians' prescriptions. With the introduction of distillation and other chemical technologies the apothecary's art became more specialized and complex, as suggested by this depiction of a distilling oven with conical sheet metal condenser heads from Hieronymus Brunschwig's *Liber de arte distillandi de compositis* (1500). Wellcome Collection, London CC BY.

chemical processes that seemed to give investigators access to deeply hidden secrets of nature. The idea that laboratory instruments might give access to the secrets of nature was revolutionary. No Scholastic commentator would ever have conceived of such a principle. Medieval alchemists espoused such views, but alchemy was essentially banished from the territory of Scholastic science. Yet practically everyone could see the extraordinary transformations that occurred in the distillation process, and they could witness transmutation anywhere perfume or *aqua vitae* was made. From plants mashed to a pulp and mixed with wine or water, an almost heavenly liquid or essential oil could be extracted, pure and immaculate. Distillation and allied technologies demonstrated the power of artisanal knowledge.

THE *AQUAVITAE* BROTHERS

The early modern was an era of new diseases. Between 1347 and 1600, Europe was struck by a wave of new and baffling epidemics. Besides the epidemic spread of bubonic plague, Europe suffered a succession of new infectious diseases, including typhus, syphilis, and virulent smallpox. Competition among healers to find new medicines to fight new contagions was a major impetus to the growth of distillation industries. Until the foundation of public health boards in the sixteenth century – and for a long time thereafter – the struggle against epidemic diseases was waged principally by the religious orders. Indeed, most of the charitable hospitals that cared for victims of the epidemics were founded by religious orders.

One such order was the Jesuati. Founded in 1360, the order was dedicated to caring for victims of the plague. Like other religious orders, the Jesuati established hospitals, visited the sick, and dispensed medicines to the poor. But what makes the Jesuati stand out among other orders is that they specialized in making distilled elixirs and cordials, which they believed preserved the body from corruption and putrefaction. Hence the Jesuati acquired a new name among the populace: the *aquavitae* brothers – from the *aqua vitae* (water of life) they manufactured and distributed to the people.

The medical doctrine underlying this practice was inspired by the writings of the fourteenth-century Franciscan friar John of Rupescissa, who taught that spirit of wine (alcohol) was the incorruptible "fifth essence" of substances, related to the four qualities as heaven is to the four elements. Following John's lead the Jesuati set up distilleries throughout Italy and manufactured remedies, which they distributed gratis to the poor. Even if *aqua vitae* didn't cure the plague, it may have enabled its victims to better tolerate their suffering.

The Jesuati's alchemical activity is documented in a manuscript composed by Brother Giovanni Andrea di Farre de Brescia between 1536 and 1562 and titled *Libro de i secretti e ricette (A Book of Secrets and Recipes)*.[1] Lavishly illustrated with figures of distillation apparatus and containing detailed descriptions of distillation procedures, the text gives prescriptions for hundreds of medicines for a wide array of ailments. Much is devoted to *mal francese* (syphilis) and there are repeated references to guaiac wood (*lignum vitae*), a famous drug from the New World. The Jesuati order spread to cities throughout Italy in the sixteenth century, where they manufactured *aqua vitae* to distribute to the poor. Other religious orders created similar distillation laboratories. Around 1360 the Dominican chapter house at the church of Santa Maria Novella in Florence established its *Officina Profumo-Farmaceutica di Santa Maria Novella*, a distillery for making perfume and medicine. The pharmacy still exists and has a flourishing perfume industry, though today it caters mainly to curious tourists.

Benedictines and Carthusians also distilled *aquae vitae*, giving brand names to modern cordials and liqueurs.

CHEMISTRY AND CLAMOR IN THE PHARMACIES AND MARKETPLACES

Apothecary shops were important sites of chymistry; pharmacy was among the earliest trades to appropriate the technique of distillation. Venice, a capital of the European trade in pharmaceuticals, had more than fifty apothecary shops in the mid-sixteenth century – about one for every 3,000 Venetians (Palmer 1985). The pharmaceutical trade thrived on the novel and exotic raw materials that poured into the city from near and far, including Mexican cochineal, bezoar from Turkey, and serpents for making the costly preparation known as theriac. Pharmacists' inventories were extensive, including dried medicinal herbs and countless gums and resins, including mastic, balsam, gum arabic, tragacanth, and dragon's blood, a resin obtained from the *Dracaena* tree. Pharmacists also manufactured and sold painter's pigments made of cinnabar (mercury sulfide), litharge (lead oxide), and other minerals.

Pharmacies were also places of sociability and leisure, spaces for gathering, conversing, and exchanging news (De Vivo 2007). All sorts of people congregated at the pharmacies – physicians, painters, tourists, patricians, and soldiers – making them favorite spots for meeting and discussion, including talk about heterodox religious doctrines. And as traders in long-distance commerce, apothecaries were sources of information and gossip about distant countries and peoples. Pharmacists didn't just tout wonder drugs made from exotic ingredients, they told stories about them. They also produced everyday drugs by distillation and other chemical methods. They sold perfumes, oils, extracts of herbs and flowers, pigments, cosmetics, *aqua vitae*, and a variety of other alcoholic beverages.

The distillation of alcoholic spirits spread roughly along the same path as the plague, partly as a result of the search for panaceas to combat the new threat. In the decades after the Black Death, physicians promoted alchemical cures, especially when they contained potable gold – prompting the observation that the age of the Black Death was also the age of golden remedies (Crisciani and Pereira 1998). It was not just the pharmacists nor "lone prophets" of iatromedicine such as Paracelsus and his followers that made and sold drugs. A host of freelancers and charlatans crowded the market squares competing to sell their electuaries and nostrums. A busy, largely unregulated medical marketplace and a thriving demand for "specifics" drew them to the piazzas, where they sought to outdo competitors by pushing their drugs under catchy trade names and making ever more audacious promises for their cures (Eamon

2006). Printing heightened the demand for panaceas, as pamphlets and leaflets proclaiming the virtues of the latest wonder drugs poured off the presses.

While apothecaries were the first tradespeople to distill alcoholic drinks on a large scale, laypeople also took up the trade. In Germany, the aquavit distilleries (*Wasserbrennereien*) began as home industries run by women. So many women confectioners and housewives were brewing brandy (*Branntwasser*) in sixteenth-century Nuremberg that the city council passed an ordinance against the practice. Despite attempts to curb the trade, the *Wasserbrennerinnen*, or aquavit women, continued to make products in home stills. Knowledge of distilling alcohol, both as a medicine and a beverage, spread throughout Germany in the fifteenth century and expanded far to the north, reaching Scandinavia by the time of the reign of King John of Denmark and Sweden. By the sixteenth century drinking alcoholic spirits had become a social problem, prompting city governments to institute measures to curb drunkenness. By the seventeenth century distilling alcoholic beverages was a major industry. Distillers were part of a new economy that was fueled by the publication of treatises on the art and by a growing demand for luxury items such as brandy and perfume (Forbes 1948: 90–1, 102–3).

EARTH TRANSFORMED BY FIRE: MINING AND METALLURGY

The metallurgical industries flourished in the early modern period, especially in southeastern Europe and the Spanish Empire. The profitability of Central European mining and metallurgical industries enriched south German merchants, such as the Fuggers of Augsburg, originally a family of dyers. Since America was the principal source of European silver, mining and metallurgy played a prominent role in Spain's imperial economy. In fact, most innovations in sixteenth-century metallurgical chemistry came from Mexico and Peru, not Europe.

Sixteenth-century metallurgical treatises vividly describe the wide range of activities making up everyday chemistry (Biringuccio [1540] 1966; Ercker 1951). Early modern metallurgists included goldsmiths, who occupied the highest rung on the artisanal ladder, followed by silversmiths, tinsmiths, bronzesmiths, and blacksmiths, who fabricated the hammers, chisels, nails, tools, and household objects that were part of everyday life. The Italian monk and social critic Tomaso Garzoni, writing in his *Piazza universal di tutte le professioni d'Italia* (*The Universal Piazza of All the Professions of Italy*), judged goldsmithing to be a "most ingenious art" that produced handsome luxury items, but he worried that "sophistic alchemists" might contaminate the trade by producing fake objects for the market (Garzoni [1585] 1996: 782). The proximity of the metallurgical arts to alchemy bothered Biringuccio, too. "Except for the manual work," he observed, "the art of the goldsmith has a close connection with that of the alchemist because it often makes a thing appear what it is not, as is seen in setting gems, in heightening the color of gold, in whitening silver, and also in gilding things that really are of silver, brass, or copper but appear to be of gold" ([1540] 1966: 363–7). In the bustling marketplaces and piazzas of early modern Europe, where charlatans mingled with apothecaries and alchemists set up shop, it was hard to tell the difference between alchemy and craft, between genuine and fake.

In the hierarchy of metallurgy, ironsmiths were at the bottom. Biringuccio dismissed them as "crude country people." Nevertheless, he acknowledged, ironsmiths have many ingenious secrets, such as the various quenching baths they made by mixing water with herb juices, worms, and oil, which they used to temper iron to the desired degree of hardness. Etching – creating intricate patterns on iron and steel using corrosive salts and acids – was first developed in Europe in the early fifteenth century and used mainly to decorate armor and steel weapons, but by the sixteenth century the technique was so common that it was used to decorate even the meanest utilitarian objects, such as butcher's and barrel-maker's knives (Smith 1988: 10–13) (Figure 5.2).

Metalworkers' skills were portable. Goldsmiths were lured into the burgeoning printing industry, where they plied their metallurgical expertise making typeface. With the discovery of rich silver lodes in Bolivia and Mexico, America also offered opportunities in metallurgical trades. Bartolomé de Medina, a tailor from Seville, was a clever inventor who saw his chance when he learned about the high costs of traditional methods of extracting silver in Mexico. Medina used silver and gold in making his dresses, and he had an idea for a cheaper extraction method. Leaving his family behind, he embarked for Pachuca, Mexico. When Medina's method, called the "patio process," was put into use in mining operations in Mexico and Peru, silver production rose steeply, precipitating a worldwide monetary crisis.

COLOR-MEN: CHYMISTRY AND THE DECORATIVE ARTS

The chymical trades associated with making colors such as dyes and painters' pigments experienced phenomenal growth and revolutionary change as a result of changing tastes and the introduction of new materials. The textile industry, which included a variety of specialized chymical workers, grew sharply. In Venice alone, the silk industry employed 25,000 workers in 1525 and more than 30,000 in 1561, while in Genoa, 38,000 people worked in the silk industry in the 1570s (Mola 2000). Dyers used a variety of vegetable and mineral dyes, including madder and kermes for red, woad and indigo for blue, and yellows from a variety

FIGURE 5.2 Chymical technologies spread. Etching was first used to decorate armor and steel weapons, but by the sixteenth century was so common that it was also used to decorate everyday objects, such as butchers' and barrel-makers' knives. This cooper's knife was made in Germany in 1702. Victoria and Albert Museum, London.

of sources. The colorant industries were revolutionized with the introduction in the mid-sixteenth century of cochineal, a colorant derived from a scale insect (*Dactylopius coccus*) native to Mexico that lives on nopal (prickly-pear cactus). Cochineal yielded a brilliant, intense, and durable red dye, far surpassing kermes (*Coccus ilicis*), a similar European insect that was the traditional source of red dye. The color red was highly prized, and cochineal quickly displaced kermes, and to some extent also madder, as a source of red. A wide range of colors –

red, scarlet, orange, purple, and black – were derived from cochineal, as dyers, painters, and apothecaries experimented to bring out cochineal's astonishing range of hues (Greenfield 2005; Padilla and Anderson 2015).

Though we don't usually think of painters as chemists, they were in fact experimenters in making and deploying colorants and pigments. As we learn from the manual of the Florentine painter Cennino d'Andrea Cennini, *Libro del'arte* (*Book of Art*, ca. 1390), Renaissance painters manufactured an array of colors derived from minerals and organic materials, many made "alchemically," as Cennini put it. Some materials, such as lac, Cennini advised buying ready-made because they were too troublesome to make. Painters could purchase the raw materials for making colors from apothecaries and *vendicolori* (color vendors), who mixed and sold colors specifically for painters and dyers (Matthew 2011). From the mid-sixteenth century, increasing numbers of painters were using cochineal to make myriad shades of red and purple pigments, thus participating as important players in the great cochineal revolution. Color was a vehicle that provided a social ladder to ambitious artisans in the chase after more vibrant and permanent colors. One of them, the Florentine inventor Cosmo Scatini, was granted a patent for high-quality black silk dying, enabling him to enroll in the dyer's guild of Venice, where he prospered (Belfanti 2004).

CHYMISTRY AND THE DOMAINS OF WOMEN

Women participated in the early modern chymical economy, though their sphere of practice was restricted. With the growth of the industrial economy, the traditional occupational structure, in which household production was integrated with the broader market, steadily declined. Since they could no longer compete for work in skilled trades, women were relegated to the margins of economic production. Nevertheless, women did practice chymical arts, most notably in the palaces and manor houses of wealthy families. Lady Isabela Cortese, the alchemical adept we met at the beginning of this chapter, was one of them, and her *De secretis* reveals a woman deeply experienced in the chymical arts.

The "stillatory," made of copper, tin, or earthenware, was a standard piece of equipment in manor houses. Some mansions had a separate still room overseen by the lady of the house. The English entrepreneur Gervase Markham, in his popular *The English House-Wife* (1651), advised the lady of the house to "have her furnish her self of very good Stills, for the Distillation of all kinds of Water, which Stills would either be of Tin or sweet Earth, and in them she shall distill all sorts of Waters meet for the health of her Household" (Markham [1651] 1675: 101). Distilled remedies and cordials were seen as a particularly appropriate remedy for the gentry (perhaps because of their tender stomachs), because distillation separated out the pure from the impure substances, thus producing a refined medicine (Rankin 2013: 105).

Ordinary women also practiced distillation and sold essences in the marketplace. In addition to the aquavit women in Germany, women distilled cosmetic and medicinal waters as a sideline to other business activities, making a living out of what has been called an "economy of makeshifts" – the patchy, improvised, desperate, and often failed strategies of the poor for material survival (Hufton 1974). Sometimes they were supported by local parishes or village welfare, but mostly they eked out a precarious living by cobbling together a variety of income-earning schemes. One of these women – her name was Maddalena – ran a little hostel in Rome and made medicinal waters and essential oils in her *bagno maria*, a simple water bath distillation vessel, similar to those used in making grappa (Storey 2011). Distillation wasn't an everyday household technology. It required specialized vessels and furnaces, which were expensive. But it was a simple enough technology that people with little education could master it. Maddalena was an experienced distiller and she used her simple chymical equipment with consummate skill.

EVERYDAY CHYMISTRY

Early modern chymistry was not a secretive alchemical practice; it was ubiquitous. If you lived in a city of any size you would see distillers in their workshops manufacturing *aqua vitae* and perfume and marvelous medicinal essences; metallurgists collecting *aqua fortis* from alembics; potters removing fired vessels from kilns; pharmacists by their furnaces mixing drugs; and countless other chymical practitioners. Although transmutational alchemists also occupied spaces in the wide panorama of early modern chymistry, they plied their trade not at the center but at the margins, and they made secretiveness essential to their art. The arcane theories some alchemists constructed to explain their art barely touched everyday chymistry. The artisan's knowledge – workshop knowledge – tended to be conservative, guided by the tried and true and oriented to practical results. Although innovation wasn't lacking, craftwork changed only slowly.

While artisanal knowledge was grounded on empiricism and the experience of generations of craftsmen, that didn't mean artisans lacked a theoretical – as it were, scientific – understanding of their work. The artisan's way of knowing the material world – "workshop science," one might call it – was radically different from the traditional medieval way of knowing. The sixteenth-century German saltpeter-maker Gerard Honrick explained the nature of saltpeter in concrete, empirical terms – a manner quite different from a typical Scholastic explanation. "The nature of saltpeter," he wrote, "is to grow in places cold and dry, where neither sun nor rain enter, nor spring resort, for the dryer and colder the place be the sooner and better do they bring forth saltpeter" (Williams 1975: 128). Saltpeter-men located the soil rich in which saltpeter grew by taste (Cressy 2013: 16). To Honrick, saltpeter was not a chemical substance in the way we might think of it, but something that "grows" in cold and moist places, just as it is the nature of rosemary to grow in dry and rocky ground. And in fact, saltpeter does grow, as microbes feed on dung to produce what we now call potassium nitrate, and saltpeter was harvested essentially as a crop when it ripened.

Artisanal knowledge of materials and material change was not theoretical or acquired from books, but learned by engaging all the senses: sight, smell, hearing, taste, and touch (Smith 2004). The potter knew his clay by how it felt in the hands, and saltpeter-men knew good sources of saltpeter by taste. This was a kind of practical knowledge that is more like cunning than science. And it was the type of knowledge that underlies the authority of master craftsmen, which, as Sennett observes, "derives from seeing what others don't see, knowing what they don't know; their authority is manifest in their silence" (Sennett 2008: 78).

Much of the culture of the workshop consists of the repetition of routine, everyday tasks. The violinist's repetition of scales sharpens hearing and sensitizes touch in the quest to perfect tone. There is in that never-ending repetition of tasks something akin to ritual, the repetition of mundane tasks in a rhythm informed by the experience of generations. Ritual suffuses the crafts as well as religion. Aside from the elaborate ritual of craft guilds, including the rite of passage from apprentice to master, there was the ritual of the work itself – think of the carpenter sanding a cabinet, a potter shaping a vessel on a wheel, or a distiller crushing herbs in a mortar to make a delicate perfume. In this culture of endless repetitions there is a search for perfection, and in that way a connection to the sacred.

CHYMISTRY IN PRINT

Although artisans typically learned about chymistry from experience in a workshop, chemical technology also spread by means of printed craft manuals, which poured from the presses beginning in the mid-sixteenth century. Roughly printed on cheap paper for mass distribution, these booklets made the tacit, unwritten knowledge of craftsmen accessible to artisans or aspiring artisans and laypeople alike. Craft manuals proliferated, first appearing in Germany in the 1530s, then spreading to the rest of Europe. These booklets were devoid of theoretical content. Instead, they spoke to immediate practical needs in an empirical voice rarely met with in scientific textbooks written for a scholarly audience.

Chemical processes made up a large share of the craft manuals. The chemical industries expanded rapidly in early modern Germany, creating a need for skilled labor. Literacy was also on the rise, and books became tools for mobility.

The German printing industry experienced phenomenal growth in the sixteenth century. By the end of the century more than three hundred printers were at work in more than a hundred German towns (Hirsch 1967: chap. 7). Economic prosperity and urban growth brought sharp increases in literacy rates, while technological change pressured craftsmen to acquire new skills, many of which could be gained or improved by reading books. Printers responded to the demand for practical knowledge with a barrage of technical and craft manuals catering to general readers.

Looming large in this sea of technical literature was a group of craft manuals known collectively as *Kunstbüchlein*, or "Skills Booklets," which appeared in various German towns in the early 1530s. Originally issued as four separate pamphlets, the booklets were best sellers: more than a dozen editions were printed within the first year and a half of publication, and the manuals were reprinted at least fifty times in the sixteenth and seventeenth centuries (Darmstaedter 1926; Ferguson 1959). Although originally written for craftsmen, their influence did not end at the workshop. The recipes making up the booklets appeared in numerous works that appealed to a broad middle-class readership, including in many of the "books of secrets" that were published throughout Europe (Eamon 1994).

The *Kunstbüchlein* first appeared between 1531 and 1532 in response to the demand for technical information fueled by Germany's expanding industrial economy. Christian Egenolff, a Strasbourg printer, brought out the first, a fifty-page booklet titled *Rechter Gebrauch d'Alchimei* (*The Proper Use of Alchemy*). The *Rechter Gebrauch* was an adaptation of a late medieval alchemical work by Petrus Kertzenmacher, said to be "a famous alchemist from Mainz." Egenolff completely changed the work, transforming it from a manual for aspiring adepts into a practical artisan's handbook. By omitting Kertzenmacher's preface on esoteric alchemy and deleting recipes that were inconsistent with the more utilitarian purposes he had in mind, Egenolff created a new kind of alchemical work, one that embraced alchemy's practical applications. For Egenolff, the "proper use of alchemy" wasn't gold-making or the search for the philosophers' stone, but practical chymistry, what today we might call applied chemistry.

The booklet's content reflected Egenolff's pragmatic approach. It included instructions for making artificial amber and pearls, recipes for etching metals, and instructions for separating gold from alloys with *aqua fortis*. It also included a section on assaying ores and making gold leaf. The *Rechter Gebrauch* contained no speculative or theoretical discussion of any kind, no discussion of transmutation, and none of the metaphorical language or cryptic symbolism so typical of the late medieval alchemical tracts. By "purifying" alchemy, stripping it of its metaphysical elements, Egenolff hoped to instruct artisans in its legitimate, practical uses (Eamon 1994: 116).

Other craft manuals published in the 1530s included several on the decorative arts, including *Artliche Kunst* (*Pretty Skills*, 1531), a booklet on dyeing, textile care, and making ink and paint for illustrating books and manuscripts; *Allerlei Mackel und Flecken aus ... zugebringen* (*How to Remove Spots and Stains*, 1532), on dyeing and caring for textiles; and *Von Stahel und Eysen* (*On Steel and Iron*, 1532), which trained readers to work with iron and steel, including recipes for tempering, hardening, soldering, etching, and coloring metals. The *Kunstbüchlein* were simple technical books for artisans and aspiring artisans, explaining chymical processes used in the household and emerging chymical trades.

Printed treatises on mining and metallurgy proliferated in sixteenth-century Europe, responding to a demand for information about the expanding metallurgical industries. The most comprehensive were *De re metallica* (1556) by the humanist Georg Agricola; the *Pirotechnia* (1540) of the Sienese mine supervisor Vannoccio Biringuccio; and *Beschreibung allerfürnemisten mineralischen Ertzt unnd Bergwerks arten* (*Treatise on Ores and Assaying*, 1574) by Lazarus Ercker, a mine master working under Emperor Rudolf II. Although the authors of these works came from very different perspectives, together their books imparted a great deal of practical metallurgical knowledge to readers. They explained and illustrated how to make assay furnaces and vessels for making various acids, such as *aqua fortis*, which alchemists, goldsmiths, and assayers all used to dissolve and separate metals. With detailed illustrations to accompany the text, the books not only explained chymical processes, but also showed them in workshop settings.

Distillation treatises abounded in the early modern period, including the great German treatise by the Strasbourg physician Hieronymus Brunschwig, *Liber de arte distillandi de simplicibus*, first published in 1500. The work, originally in German, was translated into Latin, Dutch, English, and Czech, and reprinted innumerable times. Brunschwig's book and dozens of imitations and augmentations provided valuable instruction to pharmacists getting up to speed in a new kind of pharmaceutical chemistry, and to countless novices hoping to master the new technology and enter one of the trades that the new art supported (Forbes 1948: chap. 5). Pharmacists had long benefited from trade manuals in the form of pharmacopoeias and recipe books. A discernable shift took place in manuals published in the sixteenth century. Early modern pharmaceutical manuals, like metallurgical manuals, were more descriptive, as if to survey the entire universe of the apothecary's trade, not just provide formulas. Emulating widely popular books on natural history, manuals such as Prospero Borgarucci's *Della fabrica de gli spetiali* (1567) provided a descriptive history of the trade, surveying in meticulous detail the corpus of remedies prepared in Italy's apothecary shops.

Probatum est: it has been tried and proven. These words, generously sprinkled throughout early modern recipe books, seem to suggest an ethos of openness and innovation supposedly characteristic of artisan culture. Pharmacy is often singled out as being among the most experimental of trades. But historians have read too much into these words. The much-repeated phrase *probatum est* (literally "it has been tried out") did not always mean an experimental trial to test the efficacy of a formula, much less to test or prove a doctrine or theory. More often than not the motive behind such trials was to improve the recipes or make them more marketable. Like all trades, pharmacy was conservative and resistant to novelty – which helps explain the limited therapeutic use of American plants like guaiacum and sarsaparilla beyond the treatment of *morbus gallicum*, or syphilis, an imported American disease.

In 1548 Giovanni Ventura Rosetti, an official at the Venetian Arsenal (state shipyard), published a comprehensive treatise on dyeing practices, *Plictho de l'arte de tintori* (*Instructions in the Art of Dyeing*), and seven years later another on perfumes and cosmetics, *Notandissimi secreti de l'arte profumatoria* (*Noteworthy Secrets of Perfuming*). His books were works of charity, he declared, intended to unchain the "plebeian arts" that had been "imprisoned for a so many years in the tyrannical hands of those who kept it hidden" (Rosetti 1973). Rosetti published the secrets of dyeing and perfume-making in order to promote the development of the manual arts in Venice. His commercial activities had taken him to cities throughout Italy, where he was able to observe craftsmen at work. Though Rosetti wasn't a chemist, the artisans he wrote about were. The arts of dyeing and perfumery, which Rosetti described in loving and scrupulous detail, were examples of what most early modern chemists did: they made utilitarian objects for a consuming class and experimented to improve their products in order to compete in the marketplace.

SECRETS REVEALED

While artisans typically learned chymistry directly from practice in the workshop, most laypeople learned about chemical processes from the printed "books of secrets" that circulated throughout in Europe, from libraries and bookstores to peddler's backpacks. Books of secrets simplified the process of learning technical information by employing the recipe as a means of communication. Simple, concrete, and practical, the recipe embodies the ancient way of teaching technical knowledge. The recipe imparts "how to" knowledge – the understanding of how to accomplish something in a practical setting. Books of secrets were about "secrets" in two senses. First, they published proprietary techniques and remedies that were unique in some way or composed of "secret" ingredients. They were also secrets in the sense that no one could explain their efficacy through logical reasoning or appeal to authority (Heinrichs 2012: 420). Printed in multiple editions by the tens of thousands, books of secrets were the voice of a form of knowledge – "how to" – never before accepted as scientific. The wave of recipes that washed over Europe in the books of secrets caused a shift in how intellectuals defined knowledge. Recipes certified the practical knowledge of craftsmen and heralded an age of how-to. Alongside distillers, dyers, metalworkers, apothecaries, and potters, the early modern community of chymists included a band of fervent experimenters whom the friar and social critic Tomaso Garzoni dubbed the *professori dei secreti*, the "professors of secrets" (Eamon 1994: chap. 4; Eamon 2010). Among the more than 500 different "professions" that Garzoni enumerated in his *Piazza universale di tutte le professioni del mondo*, the professors of secrets were new entrants onto the social scene of early modern Europe. In his kaleidoscopic work, Garzoni created an indelible portrait of them as a community of hypercharged experimenters who burned with such passion for secrets that "they yearn for them more than life's daily necessities" ([1585] 1996: 324). The professors of secrets were, literally, creations of the books of secrets, a new scientific identity created by the printing press. The stereotype first appeared in what would become early modern Europe's most popular book of secrets: Alessio Piemontese's *I Secreti (Secrets of Alessio*, 1555). Besides providing a potpourri of useful recipes on everything from cosmetics to metallurgy, the work was also a manifesto denouncing the practice of secrecy (Eamon 1994: 139–47). No author, however, was more successful at exploiting the new identity of professor of secret than Bolognese surgeon Leonardo Fioravanti, the self-proclaimed founder of a "new way of healing." A university dropout, Fioravanti used print and chymistry to market himself, and his sensational books made him a famous author and opened the door to the court of King Philip II of Spain (Eamon 2010). These daring and charismatic experimenters flaunted their contempt for authorities, whether of physicians or scholars, and proclaimed their adherence to empiricism and the "way of nature." Chymistry was the medium they used to skyrocket to fame. Chymical secrets became commodities and the professors of secrets cashed in.

The proliferation of craft manuals and books of secrets made technical literacy a practical possibility. They also served to popularize the chemical arts, making them less esoteric. For a growing number of people, including artisans as well as nonartisans, literacy, not membership in a guild or a formal apprenticeship, became the precondition for knowing the secrets of the arts. Technical manuals also contributed to standardization of procedures in the crafts and brought the craftsmen's laboratory – the workshop – closer to intellectual circles, enabling scientists to compare theoretical claims with technological results.

THE VICISSITUDES OF PATRONAGE

While most early modern chymists made a living from a craft or, like the professors of secrets, from the sales of their books, a few were able to climb to the top of the social ladder by gaining entry into a princely court. Employment in a court was not like having a normal occupation. Although the potential for riches was practically limitless, there was no job security. One was always subject to the capricious will of a prince or monarch. Lured by the potential for riches, chymical practitioners found their ways into courts in impressive numbers. Alchemical laboratories were set up in courts throughout Europe, including one in Florence established by the Medici Grand Dukes depicted in an engraving by Giovanni Stradano. The illustration, which appeared in Stradano's great portfolio of prints, *Nova reperta* (*New Inventions of Modern Times*, 1590), is a representation of the *fonderia*, or laboratory, that Grand Duke Francesco I d'Medici constructed in the Uffizi palace in Florence (Beretta 2014). The engraving portrays distillers, metallurgists, glazers, and other workmen busy at various alchemical tasks (Kieffer 2014). In embracing alchemy, Francesco d'Medici could project the image of himself as a forward-thinking prince and patron of the arts (Figure 5.3).

Such a scene was replicated in many early modern courts. Alchemy was a fashionable scientific discipline that also produced concrete and practical results in the crafts. In 1609 the German Landgrave Moritz of Hessen-Kassel, an enthusiastic patron of alchemy, established a chair of *chymiatry* (medical chemistry) at the University of Marburg (Moran 1991a). Alchemists were also busy at work in the court of Holy Roman Emperor Rudolf II (Purs and Karpenko 2016). Rudolf's fascination with alchemy was nurtured in the court of his uncle, King Philip II of Spain, where Prince Rudolph was sent as a teenager. In Madrid the future emperor would have encountered frenetic alchemical activity, as Philip, an avid devotee of alchemy, built distillation laboratories and gardens to furnish materials for making essences and hired foreign alchemists to the court to conduct experiments. By the 1580s, Philip was pouring substantial sums into his alchemical passion, including enough to construct a distillation laboratory at his retreat in El Escorial. The centerpiece of the laboratory was a gigantic *torre filosofal* (philosophical tower), a distillatory more than 20 feet high that could produce 200 pounds of essential oils and medicinal waters per day (Rey Bueno 2009). If you were a devotee of iatrochemistry, Philip's court was the place to be (Eamon 2016).

While the court was alluring, it could be dangerous territory for an alchemist or professor of secrets. In the court, everything and everyone served the prince. As the sixteenth-century poet Torquato Tasso wrote in his dialogue *Malpiglio*, illusion was the substance of court culture. It took myriad forms, from the

FIGURE 5.3 Chymistry in the courts. Engraving from the *Nova reperta* (*New Inventions of Modern Times*) by Jan Collaert, after Jan van der Straet (Stradanus), depicting workers at a distillation laboratory in the court of Francesco I d'Medici in the Uffizi palace in Florence. The illustration depicts workmen busy at various alchemical tasks, and in the foreground is a scene of a distillation experiment overseen by court alchemist Sisto de Bonsisti, who peers at a book of alchemical secrets. Grand Duke Francesco looks over his shoulder. Jan van der Straet (1523–1605). Copper engraving. The Metropolitan Museum of Art, New York.

court buffoon's sleight of hand to the passion for distorting mirrors. Such an environment might seem inviting for an alchemist, who rashly promises to alchemically convert cheap metals into gold. Unsurprisingly, sometimes they came to a bad end. When under alchemy's spell, adepts often promised more than they could deliver, as when in 1596 Georg Honhauer pledged to enrich the coffers of the cash-strapped Duke of Württemberg by claiming he could transmute iron into gold. Convicted of fraud, he was hanged from a thirty-foot gallows made from the iron the duke had provided for his alchemical experiments (Nummedal 2007). Leonardo Fioravanti, Europe's most famous professor of secrets, met his end in the court of King Philip II of Spain, where he was prosecuted for killing a patient with his alchemical "cures."

THE CHEMIST'S IMAGE

To many people in early modern Europe, "alchemist" was a term of scorn. To be an alchemist was to be a charlatan, a fool, and a fraud. Writers and painters made jokes of them, portraying scruffy alchemists in rags, disheveled laboratories with weighty tomes and exploding alembics, and impoverished and neglected wives and children. "Many [alchemists] have ruined themselves, but precious few have earned financial gain," wrote Sebastian Brant in his satirical poem *Ship of Fools* (1494). The alchemist was for Brant a symbol of the archswindler (Ziolkowski 2015: 38). Brant's *Ship of Fools* became internationally popular, spreading the image of the fraudulent alchemist. Even before Brant, Petrarch's *De remediis utriusque fortunae* (*Remedies for Fortune Fair and Foul*) had established the image of the alchemist as a fool and victim of a hopeless delusion. Alchemy transforms everything but the metal, Petrarch wrote: families break up, patrimonies disappear, and alchemists themselves are deformed by their foolish obsession. The image of alchemy as an insane obsession persisted throughout the sixteenth century. Erasmus, who had probably read Brant's *Ship of Fools*, described alchemy as dangerous and bewitching, "a disorder so intoxicating that once it strikes a man it beguiles even the learned and prudent" (Figure 5.4).

Some alchemists used the figure of the *Betrüger* (swindler) as a foil to promote a more positive image of the alchemist. In the process, alchemical authors in the late sixteenth century cultivated a radically different persona: the honest alchemist-expert who could navigate the alchemical marketplace and distinguish real alchemists from frauds. The German physician and alchemist Leonhard Thurneisser, for instance, utilized this strategy in his treatise on Paracelsian medical alchemy, *Magna alchymia* (1583). Thurneisser's book was in effect a tutorial on distinguishing true alchemy from alchemical fraud (Nummedal 2007: 65).

Although alchemical fraud was real, ironically the public image of the fraudulent alchemist also worked, in a cultural way, to define what an alchemist was *not* (Nummedal 2007: 6). While the suspicion of alchemy as fraud ever lurked, many extolled alchemy as serving important social needs. Worries over fraud perhaps mirrored worries about a competitive marketplace that overlooked dishonesty in commercial dealings. Even legitimate chymical arts seemed to blur into alchemy, as Biringuccio perceptively observed. Biringuccio identified two different pathways to alchemy. One was "the just, holy, and good way" that artisans follow. The other was the road taken by fraudulent alchemists, who cheated and robbed people. "It is an art founded on appearance and show, containing only vice, fraud, loss, fear, and shameful infamy" (Biringuccio [1540] 1966: 336). The *Rechter Gebrauch d'Alchimei* made a similar distinction

FIGURE 5.4 The alchemist's image. This painting, titled *The Alchemist's Experiment Takes Fire*, by the Dutch painter Hendrik Heerschop, depicts a flask exploding while the alchemist is doing a transmutation experiment. The alchemist's tattered clothing (and that of his family in the background) is meant to show the futility of the alchemist's quest. Images depicting the fraudulent alchemist were common in the sixteenth century. Science History Institute, Philadelphia.

between proper and illegitimate alchemy: not a scientific but a moral distinction. "Proper" (*rechte*) alchemy – the alchemy of the artisan's workshop – had honest intentions and produced useful results; illegitimate alchemy was nothing but false promises.

By contrast, the chymical arts in the world of artisanal crafts were held in high esteem. Garzoni praised dyeing as a noble art, but he warned of the danger of fugitive colorants. He extolled distillation as a "divine art," but he fretted that scoundrels debased the art by selling useless "wonder drugs" to gullible buyers. And while he regarded goldsmithing as a "most ingenious art," he worried about fake objects in the marketplace ([1585] 1996: 769, 782). The marketplace was susceptible to fraud. Chymical industries were particularly vulnerable because they all (as Biringuccio said) "make a thing appear what it is not" – the very definition of fraud. The city fathers of Venice worried so much about fake and fugitive colorants in its lucrative dyeing industry that they created an agency to regulate silk dyers, the *Provveditori alla Seta*. The new agency sometimes clashed with dyers over novel colorants, such as Mexican cochineal (Mola 2000: 112–13). In the popular imagination, alchemical fraud shaded into fraud in the marketplace. Many people in early modern Europe accepted the basic principles of alchemy, even the transmutation of metals and the possibility of turning base metals into precious metals. Many had seen it done in workshops. A Fugger newsletter report of January 1590 that the renowned Italian alchemist Marco Bragadino had "changed a pound of quicksilver into gold" in Venice was big news. People believed it could be done, even if rarely (Nummedal 2007: 16). Others were skeptical. Captain Gian Andrea Doria, the commander of the Mediterranean fleet, scoffed at Bragadino's braggadocio, predicting that his efforts "will soon end in smoke like that of all the others in this profession" (Goodman 1988: 26). To Doria the notion of gold-making defied common sense. As a sailor and a practical man, he trusted common sense more than fanciful alchemical theories.

CHEMISTRY AND WARFARE

Chemical warfare in Western Europe was a direct outcome of the introduction of gunpowder from China. The first chemical explosive invented by humans, gunpowder was a by-product of Chinese alchemy. Although the route of transmission of gunpowder to the West is uncertain, the explosive was familiar in Europe by the thirteenth century and was first used in warfare at the battle of Crécy in 1346. Early firearms were unreliable, but cannons found their niche in siege warfare. "No wall exists, however thick, that artillery cannot destroy in a few days," Machiavelli observed (Parker 1988: 10). Machiavelli was wrong: military architects were developing a new system of defense against artillery. Nevertheless, the use of cannon in siege warfare expanded rapidly as nations competed for power and territory. During the entire early modern period, with only a few brief moments of peace, the continent of Europe was continually at war.

The chemistry of gunpowder is simple, though it took generations of experimentation by gunpowder-makers to find the optimal combination of its ingredients. Gunpowder is composed of saltpeter (potassium nitrate, KNO_3), sulfur, and charcoal in varying amounts depending on its intended use. Saltpeter was the principal component of what the Chinese called *huo yao*, "fire drug." Like bread, wine, and cheese, saltpeter owes its existence to the action of microbes. Also known as niter, saltpeter is a waste product of bacteria that feast on decaying organic matter, similar to the biochemical action taking place in a gardener's compost heap.

Under high heat saltpeter instantly breaks down, releasing large amounts of oxygen. The other ingredients in gunpowder – sulfur and charcoal – ignite using that liberated oxygen and create the sound and fury of thunder, as contemporaries invariably described it. Everyone recognized that saltpeter was "the mother of gunpowder," but the exact ratio of ingredients in gunpowder was uncertain. How much saltpeter to include in the mixture was something that had to be worked out through long trial and error. Recipes for making gunpowder suggest continual experimentation with varying amounts of ingredients and the constituent substances themselves.[2]

Manufacturing gunpowder required reliable supplies of raw materials. Charcoal was abundant, produced from woodlands throughout Western Europe. Sulfur, also known as brimstone, could be easily obtained from the volcanic regions of southern Italy and Sicily, and in some parts of Europe from mineral springs. For gunpowder manufacture the supply of saltpeter was the perennial problem. Although saltpeter-rich soils are common in Asia, nitrate deposits are harder to find in Western Europe, and the scarcity of saltpeter was the major impediment to the expansion of gunpowder use in warfare.

European rulers attempted to stimulate and control domestic production through the appointment of saltpeter-men, who fanned out across the countryside and were authorized to dig for and remove waste soil from pigeon houses, barnyards, and stables. They were not a popular lot. The noted seventeenth-century chemist Robert Boyle, whose family owned an estate in Stalbridge, Dorset, complained of "those undermining two-legged moles we call saltpeter men" who attempted to excavate his grounds. "My pigeon house they are already digging up and would have done the like to my cellar and stables if I had not ransomed them with a richer mineral that they contain." The saltpeter-men – chymists who played a vital role in the gunpowder revolution – roamed the countryside in search of earth impregnated with dung and urine, and they were licensed to take it from private property. As Boyle discovered, the depredations of saltpeter-men could sometimes be offset by bribes, but the material they collected was indispensable for furnishing the state with munitions (Cressy 2013: 9).

But harvesting saltpeter from barnyards and dung heaps was never sufficient to meet the demands of warfare. Having observed where they could find saltpeter in nature, artisans attempted to create the same condition artificially by devising means to hasten the decay of organic materials. These innovations developed into saltpeter plantations where saltpeter was manufactured on an industrial scale. A manuscript dating from 1561 by Gerard Honrick, a German who sought to profit by establishing a saltpeter works in England, described how to make niter beds. Several niter beds were constructed in England according to Honrick's directions, including one in Colbury of 45,000 square feet. Once deployed on a systematic basis, Honrick's niter beds – essentially gigantic compost heaps – enabled England to become self-sufficient in saltpeter by the end of Elizabeth's reign (Williams 1975).

Honrick's method of saltpeter manufacture illustrates the keen empiricism that characterized everyday chemistry. He stipulated that "black earth" (composted wastes and fecal matter), urine "of persons which drink either wine or strong beer," dung "of horses which be fed with oats," and lime from oyster shells be mixed and shaped into piles. The piles had to be turned fortnightly for a year. Honrick's recipe describes the materials and conditions that modern chemists confirm as being required to make saltpeter. Even the details ring true: for example, Honrick's stipulation of urine from wine and beer drinkers is explained by the fact that ammonia levels in urine increase as the body metabolizes alcohol. Honrick's advice to use the dung of oat-fed horses probably represents a means of enhancing potassium (oats are high in potassium), which reacts with nitrates to form saltpeter that is less hygroscopic (less susceptible to the damaging absorption of moisture) than other salts. Keeping one's powder dry was every gunner's first order of business (Hall 1997: 75–6). Honrick had no theory to guide him in making these choices; it was trial and error, as improvisational as a jazz riff, performance done in an entirely new art, with little precedent and without tradition to serve as a guide.

Natural saltpeter is too crude to be used for gunpowder. Saltpeter earth had be refined, a laborious process that could take weeks. Lazarus Ercker, in his *Treatise on Ores and Assaying*, describes in detail the time-consuming process by which crude saltpeter earth was refined by multiple stages of lixiviation, or leaching, boiling, condensing, and crystallizing. Once collected, saltpeter earth, ideally dry as dust, had first to be leached with water to separate out calcium and magnesium salts that would otherwise weaken the saltpeter and make it more prone to absorbing moisture from the atmosphere. After additional boiling and leaching, the substance was dried to form crystals of potassium nitrate. The entire process could take a week or more, as workmen filled giant cauldrons with boiling water and saltpeter and emptied them over cloth filters to strain the saltpeter crystals from the impregnated earth (Partington 1999: 314–15).

Saltpeter-men didn't learn their methods from books such as Ercker's. They plied their trade without reliance on philosophical knowledge or technical manuals. They learned the craft in the traditional way, by doing and seeing, and by following the directions of master craftsmen. They were perceptive empiricists who practiced with little guidance from scientific theory. Instead they followed time-tested workshop rules and principles. Unsurprisingly, saltpeter-men are among the forgotten early modern chymists. Like gunpowder-makers, most of the people responsible for the innovations in saltpeter production are anonymous. The pathway of innovation in saltpeter manufacturing is almost impossible to trace. It was even in its own day a mysterious trade. But its importance to the economies and defense systems of Europe can hardly be overestimated. Without gunpowder's principal ingredient, saltpeter, there could be no gunpowder munitions.

Gunpowder-making, a new chymical craft, attracted adherents from across Europe: alchemists, blacksmiths, and adventurous entrepreneurs. Even enterprising peasants got into the trade. Gunpowder-making provided an avenue of social mobility for a host of ambitious men who came forward with schemes to manufacture gunpowder, men such as Francis Lee and his brothers, who grew wealthy supplying gunpowder to the British Crown, and even for immigrants such as Hans Wolf, who became one of Henry VIII's powder-makers (Cressy 2013: 64). Throughout Europe, migrant labor supplied a significant share of expertise in new trades such as powder-making and the distillation industries. In the new chymical trades, people could enter the labor market freely and move to wherever they found work. Necessarily the technology and chemistry of guns and gunpowder were matters of long-term experimentation. By the mid-sixteenth century Europeans had achieved a general consensus of how gunpowder should be made, what guns looked like, and how they should behave (Cressy 2013: 14).

Though they knew how to harness saltpeter for explosive weaponry, philosophers were hard-pressed to explain its nature. Alchemists, natural philosophers, military technicians, and minerals experts speculated on the chemistry of saltpeter, but none could fathom the way it was formed, nor could anyone fully explain its impressive effects. Scholarly knowledge of saltpeter was vague and of little practical use. Scholars knew that saltpeter was a necessary ingredient to make gunpowder, but they hadn't a clue how it was made or where it came from. Renaissance authors typically analyzed saltpeter in terms of its humoral properties – hot, cold, wet, or dry – and they puzzled over the ambiguities of a substance that shared attributes of the animal, vegetable, and mineral kingdoms. Paracelsians attributed the action of saltpeter to a vital generative principle, "a notable mystery the which, albeit it be taken from out of the earth, yet it may lift up our eyes to heaven," as the English Paracelsian Thomas Timme rhapsodized (Cressy 2013: 14). Whether saltpeter was the *miraculum mundi* or the *materia universalis* would have mattered little to those who actually made the substance. Such theories were of little help in trying to improve saltpeter manufacture.

THE GLOBALIZATION OF CHYMISTRY

In the sixteenth and seventeenth centuries American and Asian animals, plants, and minerals entered into international circulation as new foods, useful medicines, and lucrative commodities – coincidentally bringing into Europe a host of novel chemical products and processes. In most cases, knowledge of how to manufacture and use the new products depended on native knowledge. Few New World natural products illustrate the connections between chemistry, trade, and empire as well as the dye cochineal, discussed earlier in this chapter. Within a few decades of its first introduction cochineal became the most important red dye in the world until the introduction of synthetic dyes, and it was part of a vigorous global trade that extended throughout Europe, North Africa, the Ottoman Empire, and Asia. Produced in prodigious amounts and shipped to Europe through the port of Seville, cochineal was traded to northern Europe, where it was used in the thriving tapestry industry in the Netherlands and France (Lee 1951). To the east it colored the famous Venetian red velvets and silks that were traded internationally. American cochineal was loaded onto the Manila galleons that traveled between Mexico and the Philippines, then along sea routes to China where it was used to dye silk. The English even used it to dye the uniforms of the British redcoats (Greenfield 2005; Padilla and Anderson 2015).

Mexican cochineal didn't win over Europe unchallenged. Local dyers' guilds resisted the encroachment of the new colorant and in some cities tried to ban its use. Cheaper than kermes, the commonly used red dye, it was favored by dyers but distrusted by guilds and city authorities. Cochineal was banned in Genoa in 1543, though the ban was eventually lifted. After a contentious debate, the *Provveditori alla Seta* of Venice, which regulated silk manufacture, eventually approved the use of cochineal (Molà 2000: 120–31).

As cochineal transformed the dyeing industry, it also spawned scientific research. In manufacturing cochineal, Europeans relied on native knowledge and expertise. Indigenous men and women knew how to cultivate, harvest, and prepare the insects for export. But cochineal's real nature baffled natural philosophers. Spain guarded its monopoly so jealously that the dyestuff's very nature remained a mystery. Hardly anyone in Europe knew how it was grown, or even whether it was animal, vegetable, or mineral.

The English chemist Robert Boyle was fascinated by cochineal, which he judged yielded "a perfect Scarlet." In 1685 Boyle asked the Dutch microscopist

Antonie van Leeuwenhoek, by then a regular contributor to the *Philosophical Transactions of the Royal Society*, to examine imported cochineal in order to determine whether it was a plant seed or an insect. Leeuwenhoek was able to confirm by microscopic examinations that the dyestuff was an insect (Kellman 2010). Boyle, a committed Baconian, regularly visited dyers' workshops while doing research for his treatise on color. He did scores of experiments on cochineal because the dye, depending on the mordant used, could produce an astonishing range of vivid colors unlike any other dyestuff. Cochineal figured prominently in Boyle's investigations, in particular his discovery of color indicators for solution analysis. If dyers could produce different colors from a single dyestuff by the addition of mordants such as tin salts or alum, Boyle reasoned, it should be possible to use such color changes as a test for the presence of certain chemical substances, such as acids and bases. It was this insight that gave birth to chemical color indicators, the results of which are still visible in the form of our familiar laboratory litmus test papers (Eamon 1980).

Vast sums of wealth were required to keep the Spanish Empire solvent, more still to finance its wars of expansion. Spain under Philip II was chronically short of money. Little wonder, then, that Philip toyed with alchemists claiming the ability to make gold. But he quickly lost faith in *chrysopoeia* and turned his gaze toward America's abundant supply of silver, made possible with the discovery of rich silver deposits in Mexico and the Viceroyalty of Peru. Silver had been mined in New Spain since the 1520s, but the discovery of the Cerro Rico silver lode at Potosí, Peru (now Bolivia) in 1545 completely transformed New World silver production (Brading and Cross 1972; Bakewell 1983).

Mining was the first stage of a complex process by which New World silver reached Seville. Processing ore involved a series of steps leading to the extraction of pure silver from ore. The traditional Spanish method of smelting proved to be inefficient and expensive. Then, in the 1550s, a Spanish tailor named Bartolomé de Medina developed a method that revolutionized the process of silver amalgamation. Medina's method, called the "patio process," required grinding silver ore into a fine powder and spreading it out onto a paved yard, or *patio*, then mixing it with water, salt, and mercury. The mixture was left to rest for several weeks, sometimes months, while the mercury gradually amalgamated with the silver in the ore. When the amalgam was heated, mercury evaporated, leaving almost pure silver (Probert 1969). Medina's innovation enabled mercury amalgamation to be employed on an industrial scale, vastly reducing the cost of extracting silver. As a result of the innovation, silver production increased dramatically, precipitating the silver crisis of the late sixteenth century. When the patio process was introduced to Potosí, silver production jumped sevenfold between 1572 and 1592, making Potosí the largest producer of New World silver at the time.

Other innovations followed. To speed up the process, Álvaro Alonso Barba, a priest who served in Potosí for more than forty years, introduced the "pan and cooking" process of amalgamation in the early seventeenth century. Instead of mixing salt, water, mercury, and silver ore in an outdoor patio, the pan process mixed the ingredients in several shallow copper pans heated with a slow fire. The innovation reduced the time it took for amalgamation to complete from days or weeks to about fifteen hours, and also used less mercury, which was becoming a scarce metal in the New World (Bakewell 1983). As a priest trained in the Thomistic school, Barba understood metals within the framework of Aristotelian metaphysics, but he also deployed alchemical concepts (Bentancor 2007). Barba's treatise, *Arte de los metales* (1640), is rooted in the medieval Spanish tradition of practical alchemy that flourished, mostly underground, in oral tradition. His innovation illustrates the remarkable creativity of creole science (Platt 2000) (Figure 5.5).

FIGURE 5.5 A chymical revolution in the New World. The patio process of amalgamation invented by Bartolomé de Medina was put to use on a large scale in Mexico. This painting by Pietro Gualdi (1846) depicts the Hacienda Nueva de Fresnillo during silver reduction through the patio process. The dark-colored disks on the patio floor contain a semiliquid mass of pulverized, low-grade silver ore that has been mixed with mercury, ore masses that were central to the patio process, which was developed in sixteenth-century New Spain and used for approximately 350 years. Wikimedia Commons.

CHEMISTRY, THE ENVIRONMENT, AND PUBLIC HEALTH

Early modern chemical industries were virtually untrammeled polluters. Few laws governed the disposal of chemical wastes and mining residue, even though complaints about pollution were heard over and over again. The burning of coal in metallurgical industries produced massive waste and fouled the air. Ovens and forges belched black smoke, and mining wastewater was routinely emptied into rivers. Polluted water discharged in the refining of silver using the amalgamation process caused long-term environmental damage in Peru and New Spain.

The health hazards of the chymical industries were well known. Already in the 1530s, the little booklet on ironsmithing, *On Steel and Iron*, warned of the toxic fumes from the ironsmith's furnace and included instructions for making electuaries that smiths should take daily to protect themselves. "Do not make light of this," the booklet warned, "these fumes are very dangerous and harmful." We have no way of knowing how many alchemists and professors of secrets succumbed to toxic fumes from alchemical experiments. Alchemical adepts such as Leonardo Fioravanti conducted hundreds if not thousands of alchemic trials during their experimental lives, thus exposing themselves to volatile and toxic vapors. And, like all adepts, Fioravanti tested his experiments by taste.

In his treatise on mining, Agricola wrote of the particularly ghastly dangers to miners caused by breathing the stagnant air released when breaking rock by fire. In Agricola's gruesome clinical description:

> The bodies of living creatures who are infected with this poison generally swell immediately and lose all movement and feeling, and they die without pain; men even in the act of climbing from the shafts by the steps of ladders fall back into the shafts when the poison overtakes them, because their hands do not perform their office, and seem to them to be round and spherical, and likewise their feet.
>
> ([1912] 1950: 215)

Mines infected by such "malignant air" were usually abandoned. Paracelsus, who worked in a mine and in a smelting plant in Tyrol, had witnessed the occupational diseases connected with the metallurgical industry. His treatise, *Von der Bergsucht* (*On the Miner's Sickness*, 1567), described various diseases (mainly pulmonary) that miners, smelter-workers, and metallurgists suffered. Besides backbreaking work that wore down the body, miners and metalworkers breathed noxious vapors from roasting ores, releasing fumes of arsenic, sulfur, and mercury and causing a host of pulmonary and dermatological conditions.

Paracelsus' treatise was one of the first works published on occupational diseases (Moran 2019).

In the sixteenth century the expanding metallurgical industries added to the harvesting of charcoal, causing further deforestation. Agricola noted with alarm the devastating environmental impact made by the mining and metallurgical industries:

> The fields are devastated by mining operations because of the endless demand for wood: and when the woods and groves are felled, then are exterminated the beasts and birds. Further, when the ores are washed, the water which has been used poisons the brooks and streams, and either destroys the fish or drives them away. Therefore the inhabitants of these regions, on account of the devastation of their fields, woods, groves, brooks and rivers, find great difficulty in procuring the necessaries of life.
>
> ([1912] 1950: 8)

The chemical industries contributed a large share of the early modern industrial growth that accelerated pollution and deforestation. Without charcoal as fuel the chemical industries could not have flourished. Charcoal was used in virtually all the chemical industries, and if charcoal wasn't used to fire furnaces, wood was. Added to the pressure of industry, gunpowder warfare took a massive toll on Europe's forests. In the battle of Calais during the Hundred Years' War, 32,000 trees in the forest of Beaulo were felled to supply charcoal for the Duke of Burgundy's cannons, depleting the woodland for generations to come (Sumption 2017: 4: 215). It has been estimated that in France forests covered 18 million hectares in 1550, but only 9 million in 1789. In Denmark, 20–25 percent of the country was forested in 1600, but only 10 percent was forested by the middle of the eighteenth century. England in the seventeenth century was so extensively deforested that the royal navy turned to the virgin forests of New England to supply its need for ship timbers. Deforestation may have led to climate change. Around 1800 a new weather phenomenon appeared: the so-called European monsoon – not a real monsoon, but an unusual pattern in which long droughts alternated with brief, violent rains (Pomeranz 2000: 56).

Europeans lived amid vast forests in the early Middle Ages. Deforestation took place so rapidly and so extensively that by 1500 they were running short of wood for heating and cooking and facing a nutritional crisis due to the elimination of the wild game that had inhabited the forests, which since antiquity provided a staple of the European diet. Alarmed by the environmental destruction, the seventeenth-century English virtuoso John Evelyn warned, "Truly, the waste and destruction of our woods has been so universal that I conceive nothing less than an universal plantation of all sorts of trees will supply and well encounter the defect" (1776: 1–3). Already in the seventeenth century scientists were sounding the alarm about environmental degradation due in no small measure to the growth of chemical industries.

CONCLUSION: CHEMISTRY IN SOCIETY

The early modern period saw the beginnings of the expansion of chemical industries and the growing prominence of chemistry in everyday life that characterizes modernity. State investment in improving chemical industries swelled; the number of chemical patents exploded; new dyes, pigments, and chemicals were introduced; and chemical activities expanded geographically, giving way to exchanges of information and techniques that enriched mutual cultures. The chemical industries were also models for a new kind of science based on laboratory research. The alchemist's laboratory, as represented in dozens of genre paintings, was the iconic image of that new conception of science. To early modern Europeans chymistry was an escalator of social mobility. New chemical technologies such as distillation carved out artisanal spaces not represented in the guilds. Distillers had no craft union, and traditional guilds such as the grocers and spicers initially resisted innovations launched by distillation. While chymistry provided a means of social mobility for many and a source of luxury items for a growing middle class, it also contributed to environmental destruction. Once covered in luxuriant forests, Europe was by the end of the early modern period only intermittently forested. The massive deforestation of Northern Europe, abetted by energy-sapping chymical industries, led to a fuel and nutritional disaster, from which it was rescued in the sixteenth century only by the burning of soft coal and the cultivation of potatoes and maize.

The economic and environmental impacts of the growth of chemical industries was worldwide. The massive influx of silver from America, made possible by innovations in the amalgamation process, is often identified as the primary driver of the price revolution, a period of high inflation lasting from the sixteenth to the early seventeenth centuries in Europe (Miskimin 1977: 35–43). In the long run the pervasiveness of chymical activities in the towns, cities, and countryside of early modern Europe changed perceptions of the natural world (Glacken 1967). Of course, chemical change has always been part of everyday life, but the complex, specialized, and technology-based chemical industries of the early modern period, requiring novel apparatus and esoteric expertise, were something new. They awakened Europeans to the possibility of endless discovery and improvement driven by science – a way the "new philosophers" of the seventeenth century contrasted to the static and moribund Scholastic philosophy of the Middle Ages. The popular genre paintings depicting alchemists attest to the public's fascination with alchemy and with the apparatus of everyday chemistry. Everything alchemists did was replicated in artisans' workshops. In that sense, despite alchemists' attempts to keep their art secret, alchemy was no secret to anyone.

Did practical, everyday chymistry play a role in shaping the Scientific Revolution of the seventeenth century? The question reprises one of the critical issues in the history of science. The question of artisanal influence on the early modern science was first framed by Edgar Zilsel in a provocative series of papers published in the 1940s (Long 2011: 21). Zilsel argued that in the sixteenth century a new group of "superior artisans" emerged as an intellectual class. Comprising artists, engineers, instrument-makers, surveyors, and navigators, they promoted experimental methods and an appreciation of precision measurement and quantitative approaches. Because they were literate, they were able to bridge the gap between artisans and scholars, thus bringing artisanal methods and materials to the attention of the literati. To Zilsel, a Marxist, they were "the real heroes of the scientific revolution." The rise of capitalism, the decline of craft guilds, and the expansion of industry and commerce during the Renaissance pried open new opportunities for artisans to ascend to the emergent middle class. From the union of artisans and intellectuals that resulted, Zilsel concluded, modern science was born.

Zilsel's thesis, radical in its time, has in recent years been given new life, and has taken research in surprising new directions (Smith 2004). The early modern chemical industries provide considerable support to Zilsel's thesis, although with one important caveat. The social history of early modern chemistry suggests that people learned about chemical processes in more direct ways than Zilsel imagined in his intermediaries, the "superior artisans." In the cities and towns and even the countrysides of early modern Europe, chemistry was ubiquitous. You could hardly venture into the public square without seeing metallurgists, distillers, and ceramists at work by kilns, furnaces, and alembics making things and, in the process of making things, making knowledge.

CHAPTER SIX

Trade and Industry: *Chemical Economies and the Business of Distillation*

TILLMANN TAAPE

This chapter is about the economic dimensions of widespread practices and techniques that we would now identify as "chemical." A large number of early modern men and women set out to harness and transform natural materials for human benefit according to their vastly different means and understandings of matter. They mined and smelted mineral ores, forged, assayed, alloyed, and transmuted metals, struck coins, cast cannon, made gunpowder, and distilled mineral acids strong enough to dissolve precious metals and etch metal plates for printing illustrations. Others extracted dyes for the textile industry from crushed roots and insects or pounded, ground, washed, and precipitated pigments out of mineral deposits. From humble households to the courts of the nobility, the ailing, the charitable, and the profit-seeking distilled brandy, medicinal waters, and healing quintessences, while writers, compilers, and printers made and sold books on all of these topics. These practices intersected many different spheres of expertise: that of alchemy, which was itself far from being a unified enterprise, different kinds of medical practice, metallurgy, as well as various crafts and trades, from goldsmiths to purveyors of strong alcoholic drinks. They shared underlying ideas, aims, and concerns regarding the nature and composition of matter and its manipulation by human art.

Although most practitioners recognized these overlaps and continuities, they did not necessarily discuss them in the same terms, and for the most part would not have described their business as "chemical." The word "chemistry" and its cognates in other languages was increasingly used throughout the sixteenth century, but it did not refer to anything resembling a clearly bounded discipline until the eighteenth century (Newman and Principe 1998). "Alchemy" did describe a category of practices and people, but it was a capacious and contested term. Although Pamela Smith (2004: 129–53) has demonstrated that alchemy provided a framework for material changes and transmutation crucial to the practices of a variety of craftspeople, it would be putting words into their mouths to summarize such diverse techniques as distilling medicines or assaying metals under a blanket term of "alchemical" practices. For example, such eminent practitioners of mining and metalwork as Leonardo da Vinci, Vannoccio Biringuccio, and Georg Agricola did not articulate their art in those terms. While they acknowledged substantial practical overlap with what "alchemists" were doing, they distanced themselves from them and derided some of the more speculative aspects of their work. Andrea Bernardoni (2014) has argued that they staked out a field of technological production and inquiry that we can usefully – if anachronistically – refer to as "chemical arts." In this discussion I will employ this term as an artificial but value-neutral and capacious descriptor for the widely shared productive practices and ideas associated with the transformation of matter.

The material aims of the chemical arts connected them quite naturally to contexts of manufacture and economic exchange. Alchemy, for example, even when it was practiced within a context of spiritual or mystical experience of the occult, was often dependent on a relationship between practitioner and patron (Nummedal 2007: 1–15; Rampling 2012). While some alchemists held long-standing and prestigious positions at court, Tara Nummedal (2002; 2007: 75–118) has shown that there was also a more workaday kind of practical or entrepreneurial alchemist who treated alchemy as a commodity. When twenty-year-old Philip Sömmering, a German pastor, lost his position in 1555, he embarked upon years of travel, stopping here and there to acquire alchemical knowledge. He bought books, but also learned directly from expert practitioners, such as the woodcarver from Erfurt whom he paid five thaler to teach him distillation. Together with a like-minded alchemist, Sömmering signed a contract to produce the philosophers' stone for Duke Johann Friedrich of Sachsen-Gotha. When metallurgist and mint-master Lazarus Ercker suggested a similar project of metallic transmutation to Duke Julius of Braunschweig-Wolfenbüttel in 1585, his proposal was all business: he specified the promised material yield and framed the undertaking as a lucrative commercial investment for the Duke. Such practices moved physician and occultist Michael Maier in 1616 to publish a work against these "pseudo-chemists" who commodified what he saw as a noble spiritual pursuit (Nummedal 2002). In 1622, Maier was hard up, and he sent a series of letters to a minor German noble requesting money, gold, and other materials for various schemes of transmutation and medicine. Complaining that he had been repeatedly shortchanged by his publisher, he asked for an advance on his next book (Lenke 2014). For all his scorn for mercenary practitioners of the chemical arts, Maier could not escape the forces of the marketplace.

Mining and alchemy were by their nature so closely associated with material increase and the judicious exploitation of resources that they provided the language early modern people used to talk about industry and commerce at the level of territories and empires. In his *Reason of State* (1589), which would come to define a fashionable new genre by the end of the sixteenth century, political thinker and diplomat Giovanni Botero advised princes on the efficient running of their territories: like an artisan, they ought to gain knowledge and dominion over the materials of their craft. This included mining for precious ores and other treasures, but also, as Botero puts it, the aboveground mine of industry and commerce, which could yield more silver than its underground counterpart. He retheorized the role of manufacture, domestic wealth, trade, and profit through the lens of alchemical matter theory (Keller 2012; Keller 2015: 35–45). While some statesmen, such as Francis Bacon, realized the productive intersections of natural knowledge, empire, and industry, many rulers were less willing to engage with lowly commerce. Writing at the end of the seventeenth century, the alchemist-physician Johann Joachim Becher sought to integrate commerce into courtly praxis by reframing material increase, wealth, and investment in a language more familiar and acceptable to German princes: the language of material transformation and perfection through alchemy (Smith 1994). By the end of our period, the chemical arts provided compelling metaphors as well as material productivity for trade and industry.

This chapter explores the relationship between the occupations, techniques, and products of the chemical arts and the economic concerns for which they would eventually become a symbol. To maintain analytic focus while covering a range of different practices across time and space, it takes distillation as its central theme. As we shall see, distillation was thought of as a powerful and highly productive process across different early modern practices, to the extent that it has become emblematic of alchemy, pharmacy, and the manufacture of strong drinks. It intersects technical and intellectual histories of knowledge, medicine, chemistry, and manufacturing. As well as cutting across a wide range of different pursuits within the chemical arts, distillation also provides a useful social cross-analytic: it was practiced and appreciated by both men and women, the illiterate as well as the learned, from humble households to the courts of Europe's high nobility.

The perspective afforded by distillation also allows us to explore the scope and fluidity of categories such as "trade," "industry," and "economy" in the early modern period. *Oeconomia* originally referred to the household, whose significance as the core unit of society was consolidated with the rise of Protestantism. The "economy" of the household involved food, shelter, and labor, but also medical and spiritual care for its inhabitants. Craftspeople and medical practitioners produced things for a living, but they were bound by official regulations and social norms that made it impossible to understand their practice exclusively in terms of what we would now call "market forces." At the level of territories or countries, too, "economic" concerns resist a clear-cut definition. As we saw above, rulers who were interested in exploiting natural resources might be much less eager to engage in commerce, being preoccupied instead with their place within economies of favors, gifts, and kinship. Providing a window into these different spheres, distillation invites us to take a capacious view of trade and industry in the early modern period.

Within discussions of the chemical arts in relation to trade, medicine plays an important part, since it was a subject hotly contested between traditional humoral theory and proponents of new approaches to medicine that drew on chemical techniques and theories. To make sense of the wide range of different medicines and practitioners, historians of medicine often invoke the idea of a "medical marketplace" (Cook 1986; Cook 2007: 133–74). Such an account give us a lively picture of the wide range of practitioners, from physicians to itinerant tooth-pullers, charlatans, empirics, authors, and printers, all competing for clients. It has the distinction of recovering the agencies of healers as entrepreneurs and of patients as discerning customers whose buying power shaped the medical landscape. It holds a risk, however, of projecting a modern understanding of market forces onto the past, and it has been justly criticized for reducing the complex and varied modes of caring for the sick to *laissez-faire* economics, glossing over intricate social and political dynamics, regulations on medical practice, and nontransactional forms of healing (Pelling 2003: 19–37, 242f). More recent work has shown how the idea of a marketplace, properly historicized, can provide insightful analysis without unduly prioritizing or essentializing market forces (Jenner and Wallis 2007: 1–10).

This chapter shows what is to be gained by situating the chemical arts within marketplaces of medical commodities and consumption more generally, but also argues that we need to avoid defining this framework in so narrow a manner as to lose important parts of the picture. Rather than attempting to explain chemical activities within a single framework of "economic" analysis, my aim is to situate them among the varied and overlapping economies of the household, charitable healing, urban commodities, and the so-called "marketplaces" of medical practice and of print. This allows us to explore how different people understood chemical practices in relation to everyday life and livelihood, labor and resources, commerce, and the common good.

To this end, the chapter visits the workplaces, shops, and writings of all kinds of early modern distillers, largely steering clear of the most famous works and names in natural philosophy, alchemy, or medicine. It begins in the German lands, where distillation was widely practiced and where it was first codified in print. A survey of the most important commodities and their circulation is followed by a case study of the first publications on distillation by a well-read but business-minded artisan from Strasbourg, the apothecary-surgeon Hieronymus Brunschwig. His successful books mark the beginning of a wave of publications on the topic over the course of the sixteenth and seventeenth centuries, which speak widely to economic aspects of distillation and highlight the significance of print culture itself as a market for chemical knowledge. Following the fashion for distilled remedies to Venice and to London, we will explore the careers of two other author-distillers, Leonardo Fioravanti and John Hester, and their strategies for turning the chemical arts to material profit in the marketplace through claims about the novelty and efficacy of their medicines and by fashioning individual reputations. Finally, in the writings of Sir Hugh Plat, we will see how the chemical arts connected individual aims of thrift and profit to larger concerns about natural productivity and the economy of resources in Elizabethan England.

ECONOMIES OF DISTILLATION IN THE GERMAN LANDS

For all its symbolic character and association with obscure alchemical ideas, distillation meant first and foremost business to many early modern people. Familiar everyday commodities such as brandy and other distilled spirits were ubiquitous in the German lands from the Middle Ages onwards, and *Weinbrenner* (distillers of wine) were a recognized trade. Although the distillation of wine was often associated with apothecaries, this did not mean that brandy was considered first and foremost medicinal; it was also a luxury good, and often taxed as such. In the fifteenth century, the *Ratsapotheke* in Berlin, for example, held a citywide monopoly for making brandy, which was served for consumption on the premises. From around 1500, however, it faced growing competition from illicit "corner distillers" and, later in the century, from Flemish migrants fleeing the Duke of Alba's reprisals against the Dutch resistance of Spanish dominance in the 1560s. The Flemish brought with them the practice of distilling strong spirits from grain mash rather than wine. This turned out to be a very lucrative business, not least because profits could be maximized by using the spent mash as cheap fodder for livestock. By 1600, prices had dropped to the extent that Berliners were consuming more distilled spirits than beer.

In wine-growing areas, the manufacture of brandy from wine or wine lees persisted. In the Alsace region, the wine cellar of the Holy Roman Empire, the harvest was sometimes so plentiful that vintners ran out of barrels before they ran out of wine, and using the surplus to distill brandy was an economical solution. *Weinbrenner* were an important part of the urban industry in the region. In the later sixteenth century, the city of Strasbourg alone exported thousands of liters of prized Rhenish brandy each year. By the end of the century, brandy was widely and often cheaply available, becoming the common man's drink in many places (Arntz 1975: 16–68, 124–34, 168–80).

In many thriving towns, one could find another distilling trade, the *Wasserbrenner* (distillers of waters) – or indeed *Wasserbrennerinnen*, for they were often women – who produced medicinal waters in their homes. Although they have been conflated with distillers of brandy or aquavit in the literature (Forbes 1970: 90f), this was usually a separate business, using much more straightforward arrangements of pans and air-cooled "helmets" or alembics (Figure 6.1). These simple aqueous distillates of various herbs, known as distilled waters, were commonly used in all kinds of medical practice, by themselves or as an ingredient in compound medicines. They might be prescribed by a physician, but were also suited to self-help and household care. A description of their use in plain German could be found in Michael Puff von Schrick's *Von den ausgebrannten Wassern* (*Booklet of Distilled Waters*), a tiny, cheap volume that was in print from as early as 1476. The woman shown tending a still in the title illustration could be a householder distilling remedies for the family, but since the book discusses the use rather than the production of distilled waters, she may well represent a *Wasserbrennerin* who could supply them (Figure 6.2). Making and selling medical commodities might place a *Wasserbrennerin* in competition with pharmacy shops, but she could also become their trusted supplier. The account books of the *Ratsapotheke* in Lüneburg show that in 1579 alone over seventy liters of rosewater and thistle water were purchased from an unknown woman in Lübeck, nearly a hundred kilometers distant (Müller-Grzenda 1996: 105–9).

The association of women with distillation extended to the highest circles of society, pointing to another important group of practitioners. As Alisha Rankin has shown, the recipe collections and medical practice of noblewomen reflected a serious interest in the subject, but they were also bound up with gendered expectations. High-born status did not free women from traditional responsibilities as household caregivers, but often extended them to the entire estate, understood as the household writ large. In this context, distillation was especially appropriate because its purifying effect chimed with gendered ideals of purity and charity. During an outbreak of disease in 1572, Countess Dorothea von Mansfeld treated hundreds of poor people in her garden, giving out medicines for free. An expert distiller herself, she also helped other

FIGURE 6.1 Distilling furnace with an air-cooled alembic known as a *Rosenhut* ("rose hat"). Hieronymus Brunschwig, *Liber de arte distillandi de simplicibus* (1500), fol. 4r. Bayerische Staatsbibliothek.

FIGURE 6.2 Title page illustration from Michael Schrick, *Von allen gebranten wassern* (Ulm, 1498). Bayerische Staatsbibliothek.

princesses to establish their own court distilleries, among them the electress Anna of Saxony (Rankin 2013: 2–17, 93–127). Anna's own proficiency as a maker of remedies came to be widely recognized among the European nobility. When the Infanta Juana of Habsburg, the princess of Portugal, was plagued with painful and embarrassing hemorrhoids, Anna sent her a water distilled from fresh stag horns, along with a copy of the recipe (Rankin 2014). As a valued gift among princesses, both recipe and remedy mediated the kind of polite exchange that was the social glue of European court politics.

Distilling noblewomen represent a small, wealthy elite of distillers without straightforward commercial aims, but this is precisely why they were important. Giving medicines away for purposes of charitable governance or political alliance does not seem to fit with the concept of a market. And yet, Dorothea and Anna were reading printed books on distillation, sourcing ingredients and equipment, employing distillers to work alongside them, and producing medicines for their subjects as well as peers. If they do not seem to act in a market, this does not mean that they should be excluded from the conversation about the economics of the chemical arts. Instead, it demands that we widen the remit of that conversation beyond a "marketplace" narrowly defined in terms of monetary exchange or proto-capitalism.

As this overview has shown, distillation was a widespread enterprise that played a significant part in a range of overlapping economies. Making and selling medicinal waters and alcoholic drinks was the livelihood of urban tradespeople, often women, who sometimes produced and traded large amounts of distilled products. As medicines and as luxury or everyday consumables, they appealed to common folks as well as the upper echelons of society. They functioned as saleable commodities, but also as economical and appropriate material objects in gendered economies of the household or estate, charitable healing, and courtly gift exchange. To understand distillation in relation to labor, expertise, and business, we will now turn to its earliest appearance in print, namely to the *Destillierbücher* or books of distillation, which became something of a genre in their own right in the sixteenth century.

DISTILLED WATERS, QUINTESSENCE, AND OTHER BUSINESS: HIERONYMUS BRUNSCHWIG'S DISTILLATION BOOKS

It is perhaps no coincidence that the first books of distillation appeared in Strasbourg, a free imperial city in the heartland of viticulture and brandy-making. Its author makes frequent reference to the practices and equipment of the *Weinbrenner*, although he was not one of them (Figure 6.3). Hieronymus Brunschwig was a commercial distiller of a different sort. He trained as a surgeon in the local guild of barbers and bath-masters, and he also became an apothecary

FIGURE 6.3 Water-cooled still known as a "moor's head," suitable for distilling alcohol. Brunschwig, *Liber de arte distillandi de compositis* (1512), fol. 21r. Bayerische Staatsbibliothek.

with his own pharmacy shop near the fish market. As a tradesman and artisan without university education, he was an unusual author of printed books. His two works on distillation, the *Liber de arte distillandi de simplicibus* (1500) and the *Liber de arte distillandi de compositis* (1512) – for ease of reference referred to as the *Small Book of Distillation* and the *Large Book of Distillation*, respectively – were not written in Latin, as their titles might suggest, but in Brunschwig's native German. Brunschwig was the first to discuss different distillation processes in any kind of technical detail. His books speak to his extensive practical experience and skill as a distiller, providing a plethora of woodcut images of furnaces, alembics, glassware, and other equipment necessary to the art, as well as illustrations of medicinal plants to help the reader in identifying

Von distillierung

Darnach müstu haben blyhen ring in der mitten ingesencket mit fyer durch gelöchert oder groß und clein licht und schwer die mittel messig von zehen pfunden, die cleinen von acht pfunden, die grosse von vii. od. viii. pfunden, also disse figur zeigt, des glichen hültzin bretter als wyt die cappeln od der offen ist, also das das glaß durch das mittel loch gon mag. Dar nach mächerlei öffen als ich hie vnden zeige will zu brennen distillieren in dem balneü marie, vff das dz glaß nit über sich stigt mag vor der schwere des anhangende blyhes so es dar an gebunden vnd gehefftet ist.

gleser genant cucurbit, vo den türschen kolben gemacht: vonn Venedischen scherben glaß, vff das sie füer erliden mögen, deren form also ist.

Darnach müstu haben etlich gleser als dz man zwei vff eynad sturtzen mag, dere form also ist, dar in zu distillieren an der sunnen als ich in dem nünzden capitel diß ersten büchs leren will.

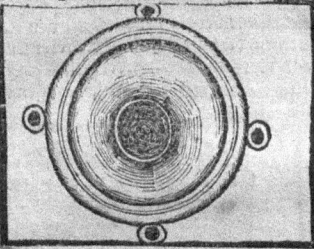

Darnach d helm vo wysser erde gebrät vnd über glasurt innen vn vssen, od kupffer, blyhyn od zynne, deren form also wurt. Ouch etlich mit zweien faltz und zwo rören, also dz der öber faltz ein röz hab, gond i die vnder röz, die helm vast vil wasser gebent.

Darnach müstu habe krüme gleser forman wie ein storck schnabel gnat retort also geformpt vnd der gleser mit zweien armen genät pelican, dere form also ist.

Dar vnder müstu haben von wysser erden wol über glasurte pfanne od blyhen od küpffern nach dynem vermögen oder begeren. Darnach müstu habe

FIGURE 6.4 Illustrations and descriptions of distillation apparatus. Hieronymus Brunschwig, *Liber de arte distillandi de simplicibus* (1500), fol. 1v. Bayerische Staatsbibliothek.

the correct herb to distill (Figure 6.4). In addition to his practical expertise, Brunschwig also emphasized his command of the theoretical, text-based aspect of distillation. Although he had no university education and only poor Latin, he claimed to have read "books great and small ... and by good masters," and as we shall see, he did engage with learned alchemical texts (Brunschwig 1500: Sig. A2v). This treatment of distillation allows us to discern how Brunschwig's concept of this process was shaped by craft practice, alchemical thought, and business expertise.

Histories of distillation as technology have tended to resort to modern-day technical and chemical concepts in their evaluation of past practices and apparatus, which tells us little about what distillers thought they were doing (Forbes 1970; Müller-Grzenda 1996). There is excellent scholarship in intellectual history that traces the trajectory of ideas about distillation in religious, medical, and alchemical traditions, but such literature largely focuses on texts, without offering a sense of how these ideas translated into everyday practice among the clutter and bustle of the workshop and the marketplace (Pereira 2000; DeVun 2009). Brunschwig's books allow us to do just that. Recent historical scholarship in the history of craft and science has outlined useful methodologies for studying knowledge associated with making, craftsmanship, and embodied experience at the intersection of ideas and practice. By challenging the dichotomy of physical work and abstract reasoning, these approaches suggest, we can uncover the world of knowledge, skill, and matter – the "material imaginary," to use Pamela Smith's term – of practitioners whose knowing and doing were inseparable (Smith 2004: 129–53; Roberts et al. 2007: xiii–xxiv, 1–7; Smith 2016). Brunschwig was one such practitioner.

The *Small Book* begins by describing distillation as a process "to separate the subtle from the coarse, and the coarse from the subtle," a common definition in medieval learned texts (Brunschwig 1500: Sig. C1r). Brunschwig's understanding of the material transformation worked by distillation was significantly shaped by his reading of medieval authorities on alchemy, in particular the fourteenth-century friar John of Rupescissa. In his *Liber de consideratione quintae essentiae (Book on the Consideration of the Quintessence of Wine)*, Rupescissa describes the quest for a universal medicine with marvelous properties. Essentially a particularly strong brandy, won through a series of sophisticated distillations, it is variously called the "quintessence of wine," the "fifth being," or "man's heaven." This substance, Rupescissa claimed, prevented matter from rotting and could cure any disease (Multhauf 1954; Pereira 2000; DeVun 2009: 60–71). Echoing Rupescissa, Brunschwig ascribes powerful effects to the process of distillation. If the refined quintessence of wine "were to move up and down the *distillatorium* a thousand times," he claims, "it would achieve the highest worth ... it becomes like heaven" (Brunschwig 1512: fol. 18r).

This process of refining "a thousand times," however, was extremely laborious, and not worth a busy craftsman's time. Brunschwig described an alternative procedure "with less effort and work, and without great cost," for those who cannot afford the "true and just way of experience … in terms of frustration or cost, but also neglect of their other business." The "true and just way" may work for those who can spare the time and money, such as "princes and lords who may achieve it thanks to their affluence," but a shortcut may be necessary for "the workman who does this with great labour, for the pleasure of princes and lords, and for his own sustenance" (Brunschwig 1512: 17r–v). It is clear that Brunschwig wrote for people who, like himself, performed distillations as part of their daily trade, to earn a living. Whether they were artisan distillers of medicinal waters, manufacturers of brandy, or itinerant alchemists hoping to impress a patron with the marvelous properties of the quintessence of wine, their distilling practice was ultimately part of a transaction that required shrewd weighing of labor and material expenses against results and rewards. They might well be unable to waste time and energy on chasing the last degree of quintessential perfection to the detriment of their "other business."

Brunschwig himself, though clearly fascinated by the quintessence of wine, in fact only devoted a few of his many chapters on distillation techniques to this marvelous panacea. The bulk of his books concerns nonalcoholic remedies such as the simple distilled waters that could be bought from a *Wasserbrennerin* or pharmacy shop. But here, too, distillation worked powerful material changes. The quintessence of wine was in fact just a very special case of a wider phenomenon: all natural things had a quintessence, and these ordinary quintessences, too, could be extracted by distillation and separated from their material dross. This meant that distilled waters did not have universal healing powers like the quintessence of wine, but specific effects on the body's four humors. As with any other remedy, one had to choose carefully; for example, to counteract a "hot" disease like a fever with a "cooling" water distilled from water lilies. Brunschwig reassuringly wrote, however, that distilled waters were much more predictable in their action upon the human body than conventional medicines. He confidently categorized them according to their effects on the humoral balance, and he told his readers how to store them in such a way that the healing power of a herb flowering in spring or summer could be kept ready for use all year round. While medicinal virtues inevitably faded over time, this was much more gradual and predictable in distilled waters compared to other forms of medicines. Even if not taken to the extremes of producing the heavenly quintessence of wine, distillation conferred some degree of the heavens' material perfection and regularity to produce mundane but reliable remedies (Taape 2014).

Since Brunschwig's waters were long-lasting but ultimately not immune to material decay, they required careful managing and storage. The two chapters dedicated to this topic remind us that Brunschwig was an apothecary and a shopkeeper as well as a distiller. Distilled waters were mostly clear liquids kept in identical flasks or earthenware jars, which could make it difficult to tell them apart if they were mixed up. It was therefore "necessary at all times to write on it the date or year of its decoction or distillation." Paper labels, in Brunschwig's own experience, were often "rotted away, gnawed at, or torn off, and I suffered great damage because of it" – he recommends engraved metal sheets instead. Protected by individual wicker baskets, the flasks should then be arrayed "according to the letters (A, b, c) to quickly find whichever you desire" (1500: fol. 11v). If this neat system was disturbed and a label misplaced or gnawed off by hungry mice, an experienced tradesman might still determine a water's freshness. A turn in color or a murky deposit at the bottom of the flask spelled the water's decay, as did the disappearance of its distinctive taste and smell. The fluid's viscosity could also provide a clue to its state, but this elusive property had to be made visible through specific tests, such as observing how it pours from a height, or how a drop placed on a thumbnail runs off. Distilled remedies were only as good as their label, and they required level-headed shopkeeping as well as the sensory expertise of the artisan who knew his products by sight, touch, smell, and taste (Taape 2014).

As well as situating his distilled remedies within economies of the shop and the workshop, Brunschwig discussed their appeal to potential patients and buyers. At the beginning of the *Small Book*, distillation is said to "make the disagreeable more agreeable." Distilled waters could be mixed into medicinal syrups or electuaries to serve them up "in an agreeable, drinkable form." Taken by themselves, they would satisfy "those who take more pleasure in a potion or water than in things that are eaten or any other medicine" (Brunschwig 1500: Sig. C1r–v). Patients could be choosy, and to anyone looking to sell or administer medicines for profit their tastes and preferences had to be paramount.

Brunschwig's sales pitch continued at the level of medical theory. Through distillation, he explained, "from the *corpus* the virtue in its soul is drawn out from each thing," such that the "soul" or healing virtue can be administered without ingesting the material *corpus* of *materia medica*. Drawing on the ancient medical authority of Hippocrates, Brunschwig explained that this was important because the "coarse" material parts of medicines could get stuck and cause damage or obstructions (Sig. C1v). By framing questions of what kind of matter should and should not be ingested with reference to its passage through the system, Brunschwig spoke to a widespread understanding of the human body in terms of fluids and motion. Learned medicine was influenced by the Hippocratic tradition of rationalizing health and disease with reference to bodily humors, good or bad, and their correct evacuations, deviations, or blockages. But laypeople, too, commonly used a language of fluids, vapors, and "fluxes" to make sense of their ailments, aches, and even emotions (Pomata 1998: 129–39;

Rublack 2002; Rankin 2008). Within this framework, Brunschwig's claims about distillation carried considerable force: it freed up the healing "soul" of medicinal substances and delivered it "to the place where it may bring greater benefit and healing," without causing harm on the way (Brunschwig 1500: fol. C1v). Although Brunschwig did not say so outright, his account of distilled waters as essentially self-directing and foolproof implied that they could safely be used for self-medication or household care by laypeople who did not have a learned grasp of humoral medicine.

In his seminal books on distillation, Brunschwig situated the manufacture of distilled remedies within traditions of alchemy and artisanal expertise, but also within economies of making, storing, and selling. He subtly theorized the commodification of medicine by chemical means, framing distilled waters as safe, efficacious, and appealing remedies for a broad social spectrum of makers and buyers. In doing so, he anticipated an increase in the commercialization of medical and (al-)chemical practices and commodities over the sixteenth and seventeenth centuries, with a growing number of business-minded individuals of various stripes making inroads into the traditional domain of regular medical practitioners.

PRINTING FOR PROFIT AND READING FOR BUSINESS: THE CHEMICAL ARTS IN THE MARKETPLACE OF PRINT

One of the advantages of an economic perspective on the chemico-medical landscape is that it invites us to view texts, and especially printed books, not simply as vessels of historical voices, but as commodities produced for a specific purpose and audience. More often than not, books were made for a market, and from the beginning of the sixteenth century onwards, these markets increasingly included socially middling readers and books to suit their needs (Fissell 2007; Rankin 2011). Alchemy, medicine, and the chemical arts represented a large proportion of these new printed works, often explicitly addressed to the "common man" of middling income and education, who could read in the vernacular but not in Latin, and whose focus was on his household and livelihood (Schenda 1982; Fissell 1992). For example, Puff's *Booklet of Distilled Waters*, introduced above, was a successful venture in the early decades of printing. Valued as a concise household pharmacopoeia of distilled remedies in straightforward German, it moved rapidly from the early folio editions to a much cheaper and more portable quarto format (Pogliani 2009). Toward the end of our period, the chemical arts had made a significant impact in the publishing world. English works alone warranted a dedicated *Catalogue of Chymicall Books*, published by the London bookseller William Cooper in 1688. It provided a census of two centuries' worth of "chymical" publications, including treatments of mineral waters, natural history, medicine, and distillation (Kassell 2011). A brief survey of the chemical arts in the marketplace of print will be a useful introduction to genres, authors, compiling and translation practices, and readers.

Brunschwig's works mark the beginning of a popular new genre of *Destillierbücher* (books of distillation) at the beginning of the sixteenth century. His two books on distillation went through twenty-three editions in the sixteenth and seventeenth centuries in German alone, and they were translated into Dutch, English, and Latin. Their wide appeal was no doubt due to Brunschwig's technical expertise, but one must also credit the ingenuity of his publisher, the Strasbourg printer Johann Grüninger. Known for its richly illustrated volumes and sloppy proofreading, the Grüninger press had a strong track record of works on natural knowledge. Brunschwig's books thus benefited from woodcut images that were at the forefront of botanical representation and from Grüninger's clever print layout that made the volumes both cheaper to produce and easier to read (Taape 2014).

Subsequent books of distillation show clear signs of Brunschwig's influence. In 1533, the physician Eucharius Röslin, better known for his writings on childbirth, published a version of a popular herbal known as the *Gart der gesundheit* (*Garden of Health*). Calling it a "noble book for the common man," he explains why he felt compelled "this winter to read it over, in addition to my other business," weeding out "many a wretched and useless thing," and adding much from his own experience: "I have no doubt that this book has been of much use in Germany, since in it is described the common art of barbers and householders, which the common people daily use for various diseases, and which the daughter tends to learn from the mother." To this "home pharmacy" for common folk and especially women's healing practices he decided to add "the book of distillation of Hieronymus Brunschwig, which is worthy of no little praise" (Röslin 1533: Sig. 1v). Upon the death of Röslin's publisher, Christian Egenolff, much of the business fell to his son-in-law, Adam Lonicer. Also a physician by training, Lonicer decided to produce a completely reworked edition of Röslin's herbal, including 300 new woodblocks, because "this book was very pleasing and sellable to the common, simple man, such that it was often re-issued in print" (Lonicer 1564: Sig. aa4r). Walther Hermann Ryff, a prolific writer and notorious plagiarist with some medical training, published his *New groß Distillier Buoch* (*New Large Book of Distillation*) in 1545, closely modeled on Brunschwig's *Small Book* with some added material, notably illustrations of more complex apparatus (Figure 6.5).

The topic received a comprehensive treatment at the hands of the naturalist and humanist Conrad Gessner, albeit under a pseudonym, in *Thesaurus Euonymi philiatri de remediis secretis* (*Euonymus Philiater's Treasure of Secret Remedies*), published first in Latin in 1552, with a German translation following

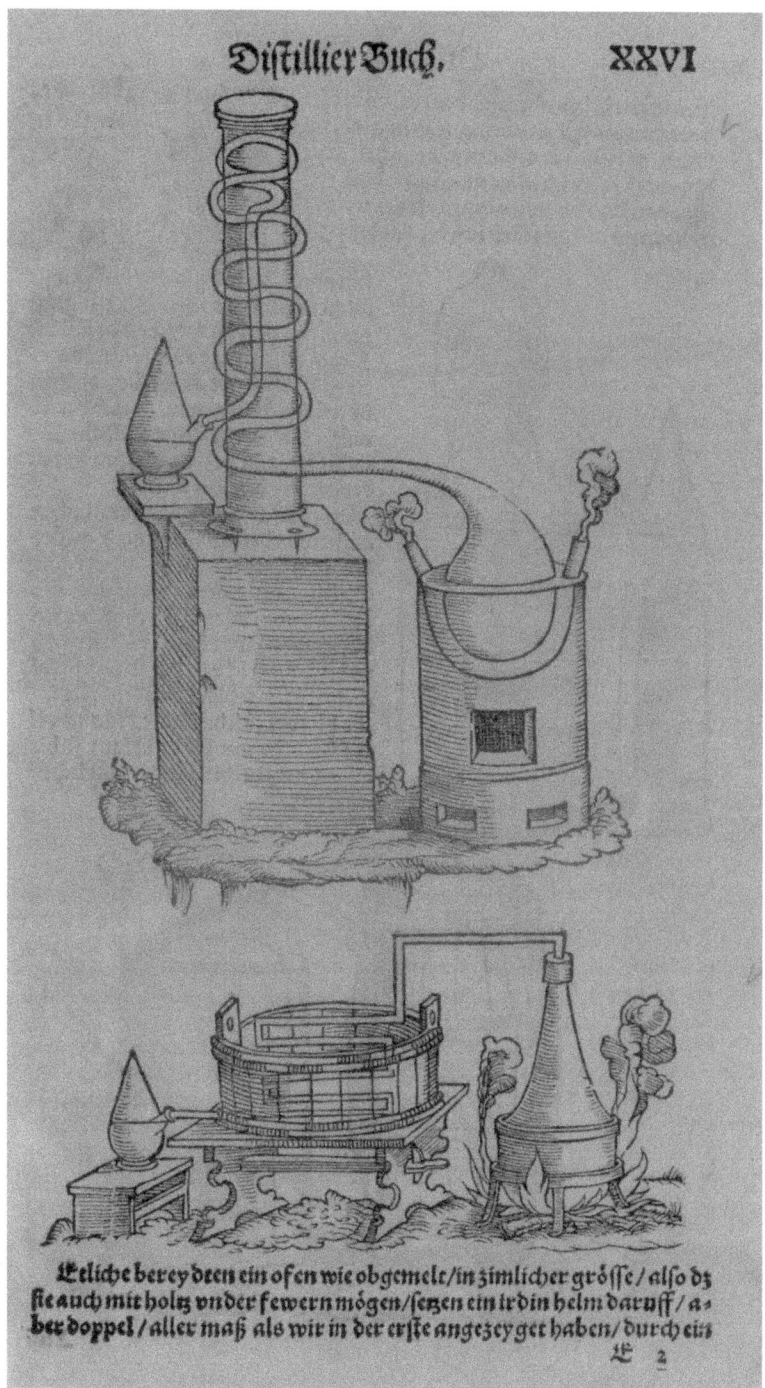

FIGURE 6.5 Elaborate water-cooling mechanisms. Walther Hermann Ryff, *Das New groß Distillier Buoch* (1545), fol. 26r. Bayerische Staatsbibliothek.

in 1555. Gessner was a well-read and diligent scholar, and he prepended to his *Treasure* a long list of the authors he had consulted in its writing, including Rupescissa, Brunschwig, "who first wrote about distilled waters in German," Ryff, and Lonicer, who had "more briefly captured the writings of Brunschwig and Ryff" (Gessner 1552: 21f, 24). Compared to these earlier writers, Gessner commanded a much wider range of learned texts. Nevertheless, he was man of practice: in a recipe for distilling *aqua fortis*, a strong mineral acid used to treat warts, he assured his readers that he had successfully "experienced and tested it with a big wart which I used to have at the tip of my finger" (Gessner 1555: 309). In early German print, books of distillation became something of a genre in their own right: busy physicians, savvy publishers, humanist scholars, and more market-oriented authors all invested considerable time and resources in this new, popular kind of book.

Outside this genre focused largely on medicine, distillation played an important role among the chemical arts associated with mining and metallurgy. Vannoccio Biringuccio was an expert in both, and in 1540 he published a practical manual on his art in Italian. *De la pirotechnia* described the whole range of activities that Biringuccio practiced and supervised as the director of the papal foundry and artillery. This included the manufacture of saltpeter, which was used in metalwork as well as for making gunpowder, but also distillation, mining, the art of the gunner, and fireworks. For Biringuccio, all of these were "arts of the fire" – technical processes that sought to work material changes and transformations. Acknowledging an overlap with alchemical practice, he nonetheless steered clear of spiritual and philosophical speculation. Writing for artisan-engineers like himself, he emphasized instead the power of his art to control natural processes by human artifice and to bring forth the "pleasing novelty" of invention (Moran 2005: 39–45; Bernardoni 2014).

In Germany, too, a number of practical treatises on the "arts of the fire" appeared, epitomized by Georg Agricola's magisterial *De re metallica* (*On Metal*). It was originally published in Latin in 1556, the year after the author's death, but this was quickly followed by a German translation in the following year. It was similar in scope to Biringuccio's work, but it dedicated several chapters to the distillation of strong mineral acids, which were crucial ingredients in metallurgy (Forbes 1970: 168–73) (Figure 6.6). *Aqua fortis* ("strong water," what we would now call nitric acid), was one way of separating gold from silver and other lesser metals, since it would dissolve the latter, but not the former. Making this *Scheidewasser* ("separating water") involved a lengthy distillation process, and once the separation was accomplished, a further series of distillations could recover not only the silver from the solution, but also much of the acid, which could be reused. These materials were costly and had to be economized, but so did the artisans' time and labor. Weighing these considerations, Agricola writes: "when making strong *Scheidewasser*, one has to

FIGURE 6.6 Distillation of *aqua fortis*. Georg Agricola, *Vom Bergkwerck* (1557), p. 366. Bayerische Staatsbibliothek.

stay awake at night, and apply exceptional diligence, labour and work." Much like Brunschwig, he offers an alternative that is less time-consuming and "not very costly" (1557: 370). Clearly and accessibly written but highly technical, metallurgical works like those of Biringuccio and Agricola were more likely to be useful to fellow professionals, mining supervisors, or artisans.

The arts of the fire also became central to ideas about medicine and the human body, notably in the works of Theophrastus von Hohenheim, better known as Paracelsus. He published little before his death in 1541, but his ideas and manuscripts were eagerly taken up by a whole host of followers who set about editing and compiling "Paracelsian" books (Debus 1977: 145–82; Moran 2005: 81–97; Kahn 2007a). Notoriously belligerent, Paracelsus called for a wholesale reform of medicine and indeed of all natural knowledge. He rejected Aristotelian natural philosophy and Galenic medicine and replaced them with his own, ostensibly new, view of the cosmos. Born in a small village in the Swiss Alps as the son of a physician and chemist, he was familiar with mining and knew how metals and minerals "grew" in the earth, were extracted by the arts of fire, and could affect miners' bodies. The human body, he later argued, could be understood entirely in chemical terms: it functioned much like a mine or an alembic, and when its internal alchemy was disturbed, the physician had to intervene by alchemical means. Distilling the patient's urine could indicate what was amiss with the inner distillations of his body and how they might be restored. Some diseases were caused by irregular depositions of salt or "tartar" and could be cured by closely related mineral substances. Controversially, many of these remedies were known to be highly toxic, but for Paracelsus, poison was a question of dosage and chemical preparation (Pagel 1958: 105–13, 126–61; Moran 2005: 67–80; Webster 2008: 140–50). This way of curing and making medicines, which he called "spagyrics," found eager followers all across Europe. Paracelsus himself, it should be pointed out, was much more a cantankerous idealist than a businessman, and many later Paracelsians, too, mainly pursued philosophical interests. Nevertheless, we shall see that their ideas and techniques contributed greatly to the rise of chemical practitioners and remedies.

Books on chemical approaches to medicine were widely read and often sold well. Johann Popp's *Chymische Medicin* (*Chemical Medicine*), published in German in 1617, combined alchemical, astrological, and cosmological ideas into medicine-making, much like Brunschwig's treatment of the Rupescissan quintessence. His three volumes, published over the course of a decade, garnered such interest among readers that several medical authors cashed in on Popp's work by producing revised and expanded versions. The last of these, Johann Helfrich Jüngken's *Clear and Well-Founded Notes on the Chemical Medicine of Johann Popp*, was published as late as 1686. Jüngken expressed concerns that doctors had lost the expertise of making medicines, and he encouraged them to "grasp the coals" for themselves (Moran 1996; Moran 2005: 49–53). As the fate of Popp's work shows, the chemical arts were enthusiastically received by medical writers and readers and lent themselves to profitable reformulations of techniques and theories of healing.

In addition to weighty tomes on technical or medical matters, the chemical arts were equally at home in more miscellaneous genres. "Books of secrets" had already circulated in manuscript form in the Middle Ages, but with the advent of print, their popularity soared. In Germany, these miscellanies of practical knowledge, craft secrets, and natural magic were known as *Kunstbüchlein* ("booklets of art"). Some, like the *Rechter Gebrauch d'Alchimei* (*Proper Use of Alchemy*, 1531), were based on works by a single author, but others were compiled by printer-publishers to satisfy the increasing literary appetite of a broad middle-class audience. *Kunstbüchlein* frequently dealt with alchemy, metalwork, distilled medicines, and other chemical arts. Although usually intended for a wider audience, books of secrets could function as vehicles of craft knowledge among tradesmen, especially in a manufacturing industry where subcontracting increasingly estranged artisans from their art (Eamon 1994: 113–26; Smith 2011). As well as informing the production of commodities, books on craft practices and secrets could contribute considerably to their enjoyment. For instance, Sven Dupré (2014) has shown that Dutch publications on glassmaking were often more about connoisseurship and appreciation of artisanal skill than actual practice.

In Italy, books of secrets were equally popular, even among the upper echelons of society. In fact, some of their authors, known as "professors of secrets," were members of the social elite rather than hands-on craft experts. One of the most lastingly popular works was called *De' Secreti del reverendo donno Alessio Piemontese* (*The Secrets of the Venerable Lord Alessio Piemontese*), first published in Italian in 1555. In one of the later editions, its author, whose real name was probably Girolamo Ruscelli, revealed that the book was informed by the collecting and testing of secrets at the "Secret Academy" he had founded in the 1540s (Eamon 1994: 134–51). As their numerous print runs, editions, and translations show, books of secrets were popular with readers and good business for publishers.

Closely related to books of secrets in their literary form, books of recipes were important vehicles of the chemical arts. Early examples, such as the *Thesaurus pauperum* (*Treasure of the Poor*) written by Petrus Hispanus (who later became Pope John XXI), mainly contained simple, affordable recipes that did not require a still or other chemical apparatus. This was an enduring concept: Brunschwig's *Large Book of Distillation* contained a section with the same title and similar contents, which went on to have a distinguished career as an independent publication, going through fifty-five editions until well into the seventeenth century. As chemical remedies gained favor across the social spectrum, they increasingly appeared in recipe collections, such as Peter Levens' *Pathway to Health*, which included "notable potions and drinkes, and for the destillinge of divers pretious waters" (Levens 1582: title page).

These printed books emerged from and remained in constant conversation with manuscript traditions. Collecting recipes – both culinary and medical – was an important part of running a household and often fell to women. Their handwritten volumes could become veritable family heirlooms, passing practical knowledge from one generation to the next and constantly changing with the addition of new recipes obtained from friends and relations, or, of course, from printed books. English memoirist Elizabeth Freke, for example, copied recipes for distilled medicines from George Hartmann's *Family Physician* (1696) and kept "stilled waters" in her medicine cabinet (Leong 2008; Leong 2013). Recipe collecting was by no means restricted to common households. As we have seen, noblewomen were keen adepts of pharmaceutical distillation, and many kept recipe books in their own hand, which might in turn be shaped by their collections of printed medical works (Rankin 2008; Rankin 2013: 67–85). Indeed, it was not uncommon for such publications to be dedicated to a princess in recognition of her expert charitable healing (Rankin 2013: 14–16). Like the remedies inscribed in them, recipes moved along gendered networks of family ties and courtly exchange, blurring the boundaries between markets and households, print and manuscript, professional and charitable healing.

Distillation was widely discussed, valued, and commodified in printed form. Effective preparation techniques, practical secrets, medical recipes, and cures all sold well, especially to an increasingly literate public of artisans and the "common man." They underpinned medical care in the household and at court or they could be read for business by artisan-distillers seeking to make a living from the production of medicines, strong spirits, or mineral acids. Instructions like Brunschwig's on efficient ways of making, storing, and managing remedies would certainly have been useful to enterprising distillers, but to attract clients, it was just as important to draw their attention to one's efficacious and desirable wares and to stand out from other purveyors. The chemical arts, as we shall see, were instrumental in framing claims about medical and technical innovation, personal fame, and expertise.

"THAT PLEASING NOVELTY": NEW REMEDIES FOR NEW DISEASES

One way of standing out among the range of medicines on offer was to emphasize what Biringuccio called the "pleasing novelty" of chemical practice (Moran 2005: 37–66). This chimed with contemporaries' perception of the early modern period as an age of the new – of discoveries, new knowledge, and useful inventions. Jan van der Straet's *Nova reperta* (*New Discoveries*) proudly displayed the most important novelties of the era: the "discovery" of the Americas, the compass, mechanical clocks, the printing press, gunpowder, and, in the bottom-right corner, a furnace and alembic for distillation (Figure 6.7). While distillation itself was not new, the chemical arts were depicted here in association with advances in manufacturing techniques and with new kinds of commodities.

With new discoveries came new challenges to established knowledge, especially in the case of medicine. In the final years of the fifteenth century, a new venereal disease spread throughout Europe, variously known as the French disease, the disease from Naples, or the pox. Scholars agreed that it was nothing like any disease described in the ancient medical texts on which they based their authority, and they were somewhat at a loss to devise a rational treatment. However, there was broad consensus that the pox had come from the New World, and soon word got around that the Spanish had brought back an effective cure from the same place (Arrizabalaga et al. 1997: 20–32). The

FIGURE 6.7 Jan van der Straet, *Nova reperta* (ca. 1590), showing the new discoveries of the age, with distillation and guaiac wood in the bottom-right corner. Courtesy The Metropolitan Museum of Art.

unprepossessing pile of logs in front of the still in van der Straet's image is in fact that remedy: guaiacum, a dark and extremely dense hardwood. It was shipped into Seville by the ton and traded throughout Europe. The well-connected Fugger merchants of Augsburg made part of their fortune by selling guaiac wood to afflicted individuals, but also to the municipal "pox houses" that were established in many cities to deal with the increasing number of sufferers (Arrizabalaga et al. 1997: 88ff; Stein 2009: 147–53). In these institutions, guaiacum was administered as a simple decoction, but it is no coincidence that the wood is piled around the still in van der Straet's image: it was not long before the chemical arts seized upon the new drug.

In the preface to his *New Large Book of Distillation*, Walther Ryff claimed that "such artful extracts made by distillation were also in part invented so that one could cure the venom and devastating evil of newly arising diseases, such as ... the French pox" (1545: fol. 1v). Other writers on distillation made the same point, claiming that traditional Galenic medicine was not up to the task of curing new diseases. In the preface to *An Excellent Treatise Teaching Howe to Cure the French-Pockes*, an English version of a Paracelsian work, the translator

John Hester (more about whom below) criticized those who sought an answer in ancient authorities, since the "the disease of the French Pocks was neyther knowne to them, nor to theyr successors," making it necessary to trust in "mens industries" such as his own thriving distilling business (Paracelsus 1590a: Sig. ϖ3r). In Venice, desperate sufferers flocked to the distillery of the "Canker Friar" Antonio Volpe in Campo San Salvatore (Eamon 2000: 202f; Eamon 2003: 126). While hidebound university doctors debated at length about the new disease and how it might be integrated into learned medicine, makers and merchants of new medicines made the most of the demand for specific and efficacious cures.

THE PROFESSOR OF SECRETS AND THE GREAT ALCHEMIST OF LONDON: DISTILLING FAME IN THE MARKETPLACE

Some purveyors of chemical medicines further bolstered their sales by fashioning a persona that stood out from the crowd of medical writers and empirics. In sixteenth-century Italy, several of the self-proclaimed "professors of secrets" gained recognition in this way, but by far the most successful was Leonardo Fioravanti, a trained surgeon and itinerant peddler of idiosyncratic cures (Eamon 2010). In 1558, he settled in Venice, a printing and publishing metropolis and a leading center of medical distillation, in no small part due to the unequaled Murano glassware, which was recognized as the best of its kind throughout Europe (Eamon 1994: 134–93; Eamon 2003).

One of the reasons for Fioravanti's success was his unerring sense for the latest medical trends, and he was quick to realize that alchemical remedies, especially distilled ones, were highly fashionable. Many sufferers were growing weary of traditional learned physic with its lengthy regimes, dietary restrictions, and weak herbal pills. Far better to take a specific drug that would effectively cure one's ailment or to try one of the universal "wonder drugs" that were increasingly on offer (Cook 1990: 411–17; Rankin 2009). Fioravanti's cures were fast, and they were powerful. Viewing the human body itself as a kind of alchemical vessel, he saw the stomach as the origin of most diseases. Accordingly, he made and prescribed chemical purges and emetics, such as his *Dia aromatica*, memorable both for its name (meaning "fragrant goddess") and its notoriously violent effect, due in large part to an ingredient we would now identify as highly toxic mercuric oxide. More than once, Fioravanti was accused of killing patients with an overdose, but his controversial approach only added to his fame. In his published works, he accompanied descriptions of his "secrets" with boastful accounts of his cures and testimonials by eminent patients. Readers were told that if they preferred not to make the remedies themselves according to Fioravanti's instructions, they could always buy them ready-made at the

"Bear" and "Phoenix" pharmacies in Venice. His medicines were still made and sold in the city until the late seventeenth century, but they also gained fame abroad. A shipment of his distilled water for eye diseases made it as far as England, and his books, too, were eagerly received in London (Eamon 2000; Eamon 2003).

Fioravanti found an eager adept in John Hester, a London apothecary, distiller, and translator of practical works on chemical medicine and surgery. Elizabethan London, bustling with apothecary-naturalists, well-traveled merchants, foreign medical practitioners, printers, and other cosmopolitan go-betweens, was a good place for anyone interested in secrets of nature and medicine and how to make a profit from them (Harkness 2007: 25–6, 142–80). Immersed in this environment, Hester came across the Italian "professor of secrets" upon the death of another compiler of secrets, Thomas Hill, who left him an unfinished draft translation of Fioravanti's plague treatise. Hester's first publication, *The true and perfect order to distill oyles* ... (1575), united Fioravanti's writings with material from, among others, Walther Ryff's *Book of Distillation* and the Paracelsian writings of Guinther von Andernach and Philippus Hermannus. In the following years, Hester completed Hill's translation of the plague treatise and published his own translation of Fioravanti's surgery as well as his *Compendium of the Rational Secrets*. He rendered Fioravanti's text almost word for word, leaving out only the occasional boastful phrase or extraneous piece of information. Tempering Fioravanti's one-upmanship and focusing on the technical detail, he sometimes produced translations that were clearer than the original, using his own expertise as a distiller to condense and clarify the text for the benefit of other practitioners (Eamon 1994: 254–6; Pantin 2013).

This did not mean that Hester passed up the opportunity to advertise his own expertise and his products in print. In his translation of Fioravanti's *Short Discourse Uppon Chirurgerie* (1579), Hester concludes his discussion of medicine making:

> if any be disposed to have any of these afore-sayd compositions redy made, for the most part he may have them at Paules Wharfe, by one John Hester practitioner in the Arte of distillation, at the sign of the Furnaises.
> (Fioravanti 1579: fol. 64r; Eamon 1994: 413)

Much like the original author, Hester lured his readers out of the book, through the city, and to his shop to buy his wares, clearly advertised by a sign showing a distiller's furnace. He even offered to teach his art in person – for a fee, of course – in a printed pamphlet hawking his remedies:

> These oiles, waters, extractions, or essence, saltes, and other compositions; are at Paules wharfe ready made to be solde, by Iohn Hester, practisioner in

the arte of distillation; who will also be ready for a reasonable stipend, to instruct any that are desirous to learne the secrets of the same in few dayes.

(Hester 1585)

A copy of this pamphlet ended up among the papers of Gabriel Harvey, and it bears a ringing endorsement in the scholar's own hand, pronouncing Hester "the great alchemist of London" (Hester 1585, copy held at the British Library; Harkness 2007: 90). Hester marketed his chemical medicines and expertise along distinctly Fioravantian lines, and the considerable success of both men relied on the chemical arts in the production of remedies, books, medical authority, and contemporary fame.

Hester was not the only chemical author-practitioner in Elizabethan London. His medical outlook was shared by a group of young surgeons – William Clowes, George Baker, and John Banister – all of whom had gained medical experience on the continent. In their publications, they confidently advocate chemical and Paracelsian cures for internal and external use. Baker, for instance, in the preface to his translation of Gessner's *Treasure of Secret Remedies*, echoes familiar *Destillierbücher* claims about the "pleasantness of the taste" of distilled remedies, and also their efficacy:

One dramme of the water of the oyle or salt of Guaiacum, for the French poxe ... Three droppes of the oyle of Cloves or Baye berries for the Cholicke: Three droppes of the oyle of Antimonium for the Leprosie, doth more than one pound of those decoctions not dystilled.

(Gessner 1576: Sig. *3v)

Baker reassures his readers that these powerful remedies are safe to take and can be easily obtained in London if "peradventure some in the sight of the furnaces, and other vessels wyll bee lothe to meddle with so busie matters," for "there be in this Citie which are most excellent in the preparing or drawing of any of them" (Sig. *3v). He recommends several of these excellent practitioners by name, including "Iohn Hester dwelling on Powles wharfe" and Thomas Hill, another cherished member of Hester's and Baker's circle of translator-distillers (Sig. *4r). London was bustling with distillers, healers, and all manner of tradespeople going about their highly skilled business, and there was one among them who made it his life's work to tap this expertise for the benefit of the country as well as his own business.

DISTILLING SECRETS FOR THE GOOD OF THE COMMONWEALTH: HUGH PLAT AND THE SCIENCE OF DEARTH

Hugh Plat was the Cambridge-educated son of a wealthy London brewer. Like Hester and the professors of secrets, he published a series of books of practical knowledge, most famously the *Jewell House of Art and Nature* (1594) – which was also the name of Plat's shop on Bethnal Green, where one could buy many of the diverse things described in the book: a lantern that would not blow out in the wind, bread that would keep during long sea voyages, and a range of distilled products used as flavoring, medicine, dye, or perfume. Plat collected most of these recipes and techniques himself, notebook in hand, among London's tradespeople, medical practitioners, and other experts. His published works only contain a small selection of the things he scribbled into his notebooks as he was looking over the shoulder of a perfumer or a cheesemaker. From John Hester he obtained a technique for catching fish in exchange for some other "secret." Seeking to make cheaper and better distilled medicines, Plat persuaded the queen's surgeon to show him his technique, and to draw him a sketch of his still into the bargain. From a metallurgist named Burchard Kranich he obtained yet another set of designs for distillation apparatus (Harkness 2007: 223–41; Mukherjee 2014: 72–84). At a glance, Plat may seem to be collecting all kinds of practical tidbits in a frantic, somewhat haphazard manner. However, he made it clear that he was not after marvelous and instructive secrets of nature for their own sake. In fact, he chided Alessio and other professors of secrets for being deliberately obscure. He insisted that their recipes had to be tested and the useful separated from the ineffective in a trial by fire that itself mirrored the chemical art of separation (Mukherjee 2011).

Far from being haphazard, there was a method and an urgency to Plat's project. When he claimed that his *A briefe Apologie of Certaine New Inventions* (1593) was intended "to the generall good of his countrey," this was no empty trope. Throughout his notes and published works, one can discern his concern with the resources of the commonwealth, material scarcity, and how it might be remedied by human art – a project that Ayesha Mukherjee (2014: 63–92) aptly terms a "dearth science." Faced with severe shortages in both hops and grain, Plat devised a way of brewing beer that required none of the former and considerably less of the latter. Everywhere he saw ingenious ways of saving fuel, substituting cheaper ingredients, and making supplies last. Distillation and other chemical techniques that could transform and preserve matter were key to this undertaking. Plat recommended distilled spirit of wine, oil of sulfur, and oil of vitriol to purify drinking water and make it last during sea voyages. The process of distillation itself was the site of many of his schemes for economizing resources. By using a larger vessel when distilling the oil of flowers or seeds one could minimize waste, and once the oil had been extracted, the remaining liquid could be recycled as scented washing water. Connecting individual thrift with the circulation and shortage of resources, Plat's dearth science intersected households, trades, and the wider economy of the commonwealth (Mukherjee 2014: 145–94).

It is owing to these wider concerns that Plat, unlike Hester, engaged more deeply with theoretical as well as practical chemical knowledge. At a time when providential Nature was apparently no longer as fruitful as she once had been, ideas about chemical transmutation were promising avenues for finding and restoring the material source of growth and plenty. In the chemical matter theories of such authorities as Paracelsus and the French artisan-philosopher Bernard Palissy, Plat encountered the idea of salt as a principle of generation. His own practical schemes included cheap and efficient ways of making bay salt from seawater, but also the use of brine and other salt-based substances as fertilizers to induce material growth in the soil. Nature was inherently generous, but it was ailing and needed human help. Not just individual trades or households but the land itself required a chemical cure to remedy its productivity. Through tireless collecting and testing of ingenious chemical practices and ideas, "nature will retourne our labour againe with an excessive usurie into our bosomes" (Plat 1594: 32f; Mukherjee 2014: 103–19). It would be anachronistic to present Plat's project as a program of economic policy, and it was never implemented as such in any straightforward way. However, historians of science have long argued that large-scale reforms of knowledge, trade, and industry, in particular Francis Bacon's utopian vision for England and the profitable science of the Hartlib circle and the Royal Society, owed much more to the expertise of artisans and tradespeople than they cared to admit (Zilsel 1942; Webster 1977). Deborah Harkness (2007: 241–53) has more recently argued that Bacon's program can be seen as an abstracted and unacknowledged reformulation of economic and epistemic practices already developed by Hugh Plat and the countless artisans and tradespeople whose expertise he collected.

CONCLUSION: DISTILLING ECONOMIES

This chapter has explored the place of the chemical arts in early modern trade and industry. These categories must be broadly defined if they are to do justice to the diversity of products, trades, and practitioners. Chemical practices such as distillation cut across boundaries of gender, social class, and intellectual commitment and training, as well as functioning in a range of different economies. Most straightforwardly, they were powerful means of material production. For Brunschwig and his intended audience as well as later authors of *Destillierbücher*, his art was first and foremost business, shaped by economies of labor, materials, and shopkeeping practices. As a technology of commodification, distillation epitomizes the transformation of natural resources into products for human consumption: it abstracts materials' properties almost entirely from their origin to preserve them in conveniently accessible, efficacious, and even fashionable forms. Chemical remedies and other products could lay claim to the novelty of technical invention in an age of new challenges and to the powerful effect they had on materials and the human body. Furthermore, the chemical arts provided a stage on which individual reputations could be created and fortunes made, in the marketplace and in print. Books were themselves chemical commodities, circulating in a market of writing, publishing, and reading. They were valued by householders, by connoisseurs of craftsmanship, and by craftsmen themselves, who were increasingly reading and writing for business.

However, trading and exchanging chemical knowledge seamlessly extended beyond the bounds of the strictly commercial. Distilling practices could travel in the form of written recipes or remedies to be exchanged within families and political networks. Distilled medicines circulated in noncommercial but vitally important economies of self-help or family and community care, in households up and down the social scale as well as princely courts. As the careers of John Hester and Hugh Plat have shown, chemical expertise was fruitfully bartered among tradespeople, paying pupils, translators of texts, and scientific intelligencers. Through constant observation, collecting, and sorting, Plat distilled from London's thriving cultures of practical knowledge a program for reversing the commonwealth's economic shortages by chemical means. The chemical arts provided not only fruitful metaphors and languages of material increase for the "reason of state," but also projects of real material improvement – less, however, thanks to learned advisors than at the hands of cunning artisans, thrifty householders, and inventive distillers.

CHAPTER SEVEN

Learning and Institutions: *Chymical Cultures at Courts and Universities*

MARGARET D. GARBER

During the early modern period in Europe (1550–1700), what we now call chemistry emerged from variegated schools of theory and wide-ranging practices of doing – whether analyzing, synthesizing, distilling, separating, purifying, or fabricating – some of which were formalized through two major elite institutions: noble courts and universities. Courts were especially receptive to chymical theories and practices, serving as a productive nexus between those persons promising to unleash the hidden powers of heaven's most immense forces or matter's most minute components and those expecting to comprehend, harness, and exhibit such powers. The intersection of these historical figures together with the material wealth and technologies at their disposal found several expression in courts, whether as philosophical and spiritual quests, economic ambitions, vehicles of commercial exchange, experimental investigations, or political displays of power through the decorative arts of gilding, painting, sculpture, and glassworks. Universities, on the other hand, opened only a few institutional doors to chymistry (the archaic spelling) and iatrochemistry (chemical medicine) prior to the late seventeenth century, first in Montpellier, then in the German territories, and only later in other European regions. As demonstrated in this volume, courts and universities by no means exhaust the numerous avenues through which practitioners traveled on their journeys to becoming chymical adepts, yet these two institutions offer vantage points from which to view the movements of physicians, laboratories, and textual sources that interconnected these sites of chymical learning.

Since this chapter cannot hope to cover all of the many courts and universities of this age, greater weight is given to German territories, where imperial and princely patronage energized so many chymical artisans and physicians and where university faculties admitted chymical arts into university curricula earlier than most. Despite this necessarily static grounding in space and time, our protagonists frequently traversed disciplinary and geographical divides in the wake of this turbulent period that marked the Lutheran Reformation and Catholic Counter-Reformation. In this time when political rule, alliances, and economies were unstable, those in power (and those hoping to benefit from their patronage) invested in various socio-religious and educational tools of statecraft, one of which was chymical. Lutheran reformers increasingly invested in claims that celestial forces could direct terrestrial consequences, including sociopolitical movements that might assist them to wrest control away from Catholic imperial or monarchical control. Catholic counter-reformers, too, invested in authoritative displays of power and theological investigations of nature. In their turn toward *natural* as opposed to *black* magic, adepts hoped "not only to describe the hidden forces of nature but also to control them, since the initiate who understood their powers could also apply his knowledge" (Evans 1973: 197). As artisans gained patrons by promising to help courts navigate, display, or direct potentially chymical and religious sources of natural power, their workshops gained heightened status and visibility (Smith 2004). Artisans attracted reformers, scholars, physicians, and investors to the potential financial, spiritual, and epistemic profits that chymical workshops might reap by investigating nature's forces at their deepest and highest levels.

Many early moderns believed that the body was the central locus for coalescing celestial and terrestrial powers, as it was seen as the microcosm of the heavenly macrocosm, and thereby chymically inclined physicians became well poised to apply theories of astronomy and physiology toward practices of chymical arts. Medical training, with hands-on practices of botany, anatomy, and healing, became the door through which chymistry entered university education and out of which several chymical physicians entered courts as personal physicians to dukes, princes, emperors, and their households. University botanical distilleries, some already producing mineral and metallic tonics, expanded to accommodate the laboratory operations necessary for creating chymical medicaments. Formalized university training in iatrochemistry interlaced chymical philosophies and experimental designs, eventually redrawing the boundaries of natural philosophy itself. In this chapter, I explore chymical learning at early modern institutions by focusing on chymical physicians who traveled between courts and universities and who served as the key liaisons through which textual and experiential learning circulated in this "golden age" of chymistry (Principe 2013: 107).

EUROPEAN NOBLE COURTS

Noble courts were ideal spaces for chymistry's growth, especially those embracing well-stocked libraries of chymical manuscripts and books embodying its long textual history, and often housing equally well-stocked workspaces with material resources to enable chymical investigations. In the late sixteenth and early seventeenth centuries, royal residences provided significant seedbeds for growing chymical knowledge by offering employment where intellectual expression flourished outside of traditional Galenic and Aristotelian-based university medical curricula. Consequently, courts cultivated scholarly and practical chymical innovations and nurtured a variety of chymical practitioners, some with formal university training in medicine and some without, such as craftspersons, courtiers, and nobles. Ever subject to perilous times, courts offered insecure employment, resulting in chymical communities that changed with each new generation of chymical adepts. Yet artisans, scholars, and physicians continued to cultivate shared forms of scholarly, craft-based learning (Figure 5.3).

Chymical learning at courts grew considerably between 1550 and 1700, especially within the German territories where emperors, princes, and princesses built chymical workspaces, hired chymical artisans, physicians, and practitioners, circulated chymical recipes, and enlarged libraries with contemporary and medieval chymical texts and treatises (Rankin 2013). To some extent this surge in chymical medicine and philosophy was an outgrowth from Renaissance humanists, who advocated a return to original Greek and Roman sources, and also to the book of nature as much as the book of scripture (Kristeller 1979; Copenhaver 1992). Whereas reading the book of scripture animated Lutherans, Calvinists, and other reformers in ways that altered histories of religion, reading the book of nature inspired chymical physicians and artisans of varied confessional beliefs in ways that altered histories of science.

The turn toward "reading" nature through direct sensory experience, physical manipulation, and repeated trials had grown to such an extent that, by the late sixteenth and early seventeenth centuries, physicians were exchanging and circulating this information informally through physicians' letters as *experimenta*, "medical and alchemical recipes and short descriptions of successful procedures" in Katharine Park's description (2011: 36). These developments contributed to an ongoing cultural development by natural philosophers of hands-on practices and Aristotelian forms of experience that, by the mid-seventeenth century, were disciplined further into the genre of *experiment*, a means of challenging truth claims through witnessed trials (Schmitt 1969; Dear 1995).

At the center of these efforts to circulate *experimenta* were *Leibärzte* (personal physicians of emperors, princes, royals, and other dignitaries), whose university training was most closely related to studying the natural world. Having acquired medical degrees, many physicians moved to courts where they served as *Leibärzte* or *Hofärzte* (physicians for court households); others gained professorships at regional universities. Frequently they held positions at both. University education in early modern Europe was structured by the Renaissance humanist movement, with its focus on classical literature, poetry, history, and Greek or Latin texts. Supplementing an inherited medieval curricula that included the *trivium* (grammar, logic, and rhetoric) and *quadrivium* (astronomy, arithmetic, geometry, and music) were ethics, metaphysics, and natural philosophy. Following the Reformation and Counter-Reformation, university instruction included theology as a crucial part of spiritual education at Lutheran, Catholic, and other confessional institutions. After acquiring Bachelor of Arts degrees, students could move to graduate education in medicine, law, or theology.

Students interested in pursuing advanced studies of the natural world entered medical school, which included not only human physiology and pathology, but also studies of minerology, botany, astronomy, astrology, and meteorology. Academic medicine also included practical arts of recording bodily responses to medicines; consequently, physicians recorded in their *experimenta* "the often unpredictable information that the practice might reveal" (Park 2011: 36). Having taken root during Renaissance humanist reform movements, the Neoplatonic–Hermetic philosophies that described correspondences that were believed to link celestial and terrestrial forces grew significantly in the post-Reformation era. To understand such correspondences, scholars questioned stellar effects on human bodies, and thus paid attention to the "science of the stars," usually in the form of astronomical observations, which helped join astronomy and astrology to medicine (Westman 2011: 34). Continental students traveling to northern Italian universities, like Bologna and Padua, to acquire a superior medical education were exposed to humanist philosophies, astrological medicine, and the genre of *experimenta*. The confluence of efforts to comprehend the correspondences believed to link the heavens and humans plus desires to direct flows of vitalist forces through hands-on practices coursed through this period, opening cultural floodgates to reinterpreting arts, religion, medicine, and chymistry.

As figures that transitioned between courts and universities, physicians served as noteworthy cultural liaisons, transmitting theories and translating practices and skills. Some *Leibärzte* became advisers not only on human health and disease, but also on astronomy, astrology, botany, and/or chymistry. As personal advisors they contributed to courts' intellectual cultures, becoming indispensable explicators of products and texts, including those of a mysterious traveling iconoclast who promised new cures based on Christian piety and chymistry.

Few physician–scholars studied personally with Philippus Aureolus Theophrastus Bombastus von Hohenheim, known as Paracelsus; nonetheless, his teachings permeated European landscapes and took shape through a variety of channels. Following some early chymical training in his youth from the alchemist Bishop Johannes Trithemius, Paracelsus traveled with his physician father to Villach, Austria, where he apprenticed in the Fugger mines, learning of metals, minerals, and diseases of miners as a youth (Debus 1977). After widespread travels to various universities and reading a variety of Hermetic, Neoplatonic, and chymical texts, he landed a promising (if short-lived) position as city physician and lecturer at the University of Basel in 1527 (Partington 1961; Debus 1977; Shackelford 2004). After creating an uproar by teaching in German (rather than Latin) and burning books by traditional philosophers, he traveled throughout the continent and launched a medical reformation through his writings, especially his 1541 *Paragranum* (*Beyond the Seed*) that linked medicine, astronomy, chymistry, and spiritual philosophy to Christian piety and practical experience (Debus 1998; Moran 2005). While Paracelsus' efforts certainly impacted the merging of medicine and chymistry for some physicians, it is important to realize that there was already a lengthy prior tradition of chymistry such that physicians in post-Paracelsian Europe had the options of adopting much, only parts, or even none of his philosophy.

Posthumous interest in Paracelsian philosophy and iatrochemistry increased dramatically starting in the 1570s, when a generation of scholars translated, systematized, and popularized Paracelsian ideas: Gerard Dorn, Joseph Du Chesne, Petrus Severinus, Leonhard Thurneisser, Theodore Zwinger, Daniel Sennert, Oswald Croll, and the powerful critic, Andreas Libavius (Shackelford 2004; Moran 2007; Principe 2013). Although followers had varied views of Paracelsian particulars, the general outline most broadly agreed upon included a new nonhumoral view of health and disease, a belief that celestial forces implanted their signatures into terrestrial things, and the doctrine that three basic principles comprised the physical matter of all things, referred to as the *tria prima*: salt, sulfur, and mercury. The *tria prima* expanded on an already existent Aristotelian concept in which metallic generation resulted from the agency of sulfur and mercury. By adding a third principal, *salt*, Paracelsus proposed that all property-producing matter was composed of three principles, which he analogized to a burning twig, where the flame was represented as sulfur, the smoke as mercury, and the residual ash as the salt (Debus 1977: 57). Yet the ambiguous status of the *tria prima* – whether it formed the irreducible principles, or constituted just secondary effects from manipulations, or whether these substances were present at all – would structure debates over the next century for both interpreters and critics.

Paracelsian medicine proposed that disease resulted from specifically located foreign agents that disrupt normal bodily functions. Each organ in the body was ruled by a specific *archeus*, or life force, which functioned as an "internal alchemist" that separated the pure from the impure, as, for example, when the stomach separated nutrients from poisons (Debus 1977; Debus 1998). If an internal alchemist malfunctioned in its role (e.g. if poisons remained), then the associated organ would be corrupted and disease would ensue. Expanding upon the widely circulating Renaissance belief that connected the larger cosmos, or macrocosm, to the human body or microcosm, Paracelsus claimed that astral projections connected the power of the stars to corresponding parts of earthly bodies and directed their specific functions. In addition to a Christian soul, human bodies contained the *tria prima* as physical matter, such that astral emanations that penetrated the terrestrial body left within it their signatures. As Bruce Moran described it:

> Just as everything in the macrocosm was made of the three principles of Sulphur, Salt, and Mercury, diseases of the body were also born into these three universal categories and manifested themselves corporeally as saline (for example with outbreaks of the skin), sulphurous (inflammation or fevers of various sorts), or mercurial (usually diseases associated with an excess of moisture or phlegm or bodily fluids generally).
>
> (2005: 76)

Given that everything in the universe was made of the *tria prima*, medicines made of metals, minerals, or chymicals with analogous properties to the affected organ were deemed efficacious since they could restore to the internal alchemist its function.

Chymical medicine grew significantly in this period but met with resistance at universities, where the foundation for medical training was the humoral-based medicine derived from the ancient physician Galen, which proposed that disease resulted from an imbalance of one of the four bodily humors (black bile, yellow bile, blood, and phlegm), not the *tria prima* or *archei*. Since Paracelsian ideas starkly contradicted Galenic suppositions that contraries cured (a disease caused by too much humoral blood, for example, was treated by removing it via phlebotomy), Paracelsian medicine was unwelcome at most universities. Courts, however, offered a place for physician–philosophers to merge their university medical training with the theoretically heterodox ideas of chymical philosophies and with the practical workshops of artisans, apothecaries, and chymists. The particular weight that specific courts gave to chymical texts and practices gave rise to a variety of strikingly different chymical traditions.

Prince Moritz's court at Hessen-Kassel

The court wielding the most significant influence for chymistry's reception and growth in this period lay in Hessen-Kassel, where Prince Moritz saw chymistry as a means to accomplish his confessional objectives. Hoping to build an alliance of Protestant princes against the Catholic Holy Roman Emperor, Moritz downplayed differences between Calvinists and Lutherans, a move that made his court attractive to many chymical practitioners and repellent to more conservative Lutherans. Moritz's worldview engaged a spiritual–practical chymistry aimed at comprehending astral connections between the heavens and earth, so that by manipulating the terrestrial microcosm he hoped to control powerful macrocosmic effects (Moran 1991c).

Moritz surrounded himself with a variety of chymical *Leibärzte* who were under his own direction: some explicated alchemical texts, some practiced Paracelsian medicine, some enhanced court pharmaceutical knowledge. Others conducted material practices with Moritz, who was himself, in the terms of Bruce Moran, a "prince–practitioner," one whose fervent interest in personally investigating chymistry went beyond a literary engagement in philosophy and arts. He was not alone. The Medici princes, Francesco I and his son Don Antonio, also spent lengthy days and nights conducting alchemical experiments in their ducal laboratories and *fonderia*, recording volumes of such works together with other secrets (Moran 1991c; Eamon 1994). Uniting the circle of chymist–theologian–courtiers surrounding Moritz were the influential chymical writings that appeared in the early seventeenth century under the pseudonym "Basil Valentine." These writings offered a philosophical anchor for Moritz and his physicians as they negotiated incoming waves of hopeful patronage-seekers offering medicinal panaceas and chymical recipes for the philosophers' stone. The search for the philosophers' stone merged ambitions for metallic transmutation with aspirations for a universal medicine that promised longevity and broad-spectrum curatives. Prince Moritz granted status and legitimacy to Paracelsian medicine and Hermetic philosophy, where they were "official court policy" within a specific brand of reconciled Calvinist–Lutheran reforms at the University of Marburg (Moran 1991c: 9).

In effect, Moritz's patronage of occult philosophy, with its promise of controlling powerful forces, offered a surrogate reality whereby a future Calvinist prince might wrest control from the Holy Roman Emperor (Moran 1991c: 171–76). The Calvinist court at Hessen-Kassel, with its tolerance for heterodox views, became a significant space for the early legitimation and expansion of chymical philosophies in the heady days prior to the Palatinate Elector Frederick V's defeat in the ruinous Thirty Years' War.

Political aspirations also connected Paracelsians to the French Huguenot King Henry IV by appealing to chymists' intent on prolonging their own "Golden Age" – an age of "peace and religious toleration" – in 1598 after the Edict of Nantes, and by appealing to a king who might gain power through patronage of chymical arts, mining, literature, and potential cosmological-political consequences (Kahn 2007b: 4). Henry IV inherited a few court chymists, like Du Chesne, from family courts, and added courtier poets, scholars, diplomats, and physicians sympathetic to chymistry. Although not a prince–practitioner, Henry's patronage created an environment in which works by chymical authors "Basil Valentine" and Michael Sendijov (Sendivogius) were translated and printed; Paracelsian royal physicians, including Count Jacob Alstein, David Lagneau, Theodore Turquet de Mayerne, Jean Ribit de La Riviere, and Jacques Troulliard, increased the court's international reputation as a Paracelsian stronghold, and Mayerne and Du Chesne gained continental renown for engaging in disputes with the Medical Faculty of Paris (Debus 1991; Kahn 2007b).

Connections among courts were facilitated through courtiers, physicians, and diplomats. The peripatetic Du Chesne traveled to German courts and dedicated treatises to German princes, while the French Ambassador Guillaume Ancel joined foreign diplomats at Emperor Rudolf II's court, patronizing chymical experimenters and their chymical treatise publications, such as Nicolas Barnaud, a member of chymical circles around Prague (Evans 1973). Liaisons of courtiers like Du Chesne helped exchange methods and materials among courts, as surely as the vast epistolary networks of scholars, noblewomen, and noblemen circulated chymical medicines and theories aimed at improving bodily and financial well-being (Rankin 2013). Following their patron's death, Henry's courtiers scattered to other courts in search of new patrons, spreading chymistry from France to England and Germany. The most lasting aspect of Henry's short reign was his founding of an office of spagyric medicine, lasting until 1776, and his active pursuit of practical chymistry by establishing a department of mining (Shackelford 1998; Kahn 2007b).

The Rudolfine court at Prague

At the seat of political power in early modern Central Europe, Rudolf II, King of Bohemia and Emperor of the Holy Roman Empire, raised the social prestige of the chymical arts by conferring upon them his socially symbolic authority, cultivated aesthetic taste, and vast political influence. As his extensive *Kunstkammer* (cabinet of arts and natural curiosities) reveals, Rudolf embodied the role of an extravagant collector of arts, music, and astronomical instruments. Increasingly enthused by merging arts and occult sciences, his interests moved beyond symbolic displays of power that such collections traditionally rendered toward gaining experience as a chymical prince–practitioner within his castle laboratory, a trait he shared with later Habsburg emperors Ferdinand III and Leopold I. Such experience was gained at the expense of considerable criticism. One Tuscan Ambassador to court disparaged Rudolf's "study of these arts" as being "against the decorum of a prince" because "he himself tries alchemical experiments, and … engaged in making clocks." Worst of all he has "transferred his seat from the Imperial throne to the workshop stool" (Kaufmann 1988: 7).

Arts and sciences, clock-making, and experiments emerge in this contemporary portrait of Rudolf II as activities joined by space, status, and manual labor.

The Rudolfine court became famous indeed for its workshops, which attracted artisans, scholars, physicians, and instrument-makers from the German territories, Italy, the Netherlands, England, and France who could merge art and nature in novel ways. By means of productions of cut glass and stone carving by Ottavio Miseroni, by goldsmiths like Christoph Jamnitzer, by significant instrument-makers like Christof Margraf, among others, the imperial workshops created "some of the finest pieces ever made in the entire history of 'decorative arts'" (Kaufmann 1988: 5). At least twenty-five court painters, including the Calvinist Jacob Hoefnagel, the Catholic Hans von Aachen, and the Lutheran Bartholomaus Spranger, created exceptionally fine art at Prague (Kaufmann 1988). Rudolf's was the court where the astronomers Tycho Brahe and Johannes Kepler, along with several Paracelsian physicians, found support, refuge, and often financial backing, whether they were Calvinist, Lutheran, Jewish, or Catholic (Evans 1973; Evans 1979). As a supremely cosmopolitan Renaissance patron, his religious tolerance and intellectual curiosity for unconventional ideas created a milieu in which occult sciences thrived and artisans interacted (Evans 1973; Kaufman 1988; Findlen 1997) (Figure 7.1).

FIGURE 7.1 View of the city of Prague. Etching by Johannes Wechter (Aegidius Sadeler II, Antwerp/Prague, 1606). Harris Brisbane Dick Fund 1953, 53.601.10 (72–83), The Metropolitan Museum of Art, New York.

Driven to learn the arts of separation and transmutation, Rudolf was introduced to workshops in his early years at the Escorial court of his uncle Philip II of Spain, where he may have learned arts of separation from Philip's experts who made chemical medicine at a distillery that in the 1570s was "perhaps the most lavish in all of Europe" and who facilitated the work of pharmacists and physicians (Rey Bueno 2015: 134). Philip maintained a library of Paracelsian, Hermetic, and chymical authors, thereby creating a court tradition of chymical arts that extended through Charles II's reign. In Prague, Rudolf built a laboratory, and in 1607 he hired Jacob Horcicky (Sinapius) as his imperial chymist (Prinke and Zuber 2018).

As Prague attracted artists and craftspersons, it also drew chymists, chymical physicians, occult philosophers, and other humanists into an orbit occupied not only by court patrons, but also by court satellites of Bohemian aristocrats and nobles. Unlike Hessen-Kassel, where Prince Moritz's patronage was crucial for the growth of chymical arts, in Prague, imperial physicians and chymists embarked on projects with little formal imperial support, resulting in dissimilar chymical cultures. As divergent practitioners of chymical arts grew, they creatively intersected with other artisans. The imperial printmaker Aegidius Sadeler, for example, engraved the frontispieces for both Croll's 1609 Paracelsian *Basilica chymica* (*Royal Chymistry*), a volume on religion, chymistry, and medicine that went through eighteen editions (Figure 7.2), and the *Leibärzte* and lapidary Anselm Boethius de Boodt's *Symbola divina et humana* (*Divine and Human Symbols*, 1601), a magnificent collection of *imprese* – symbolic and mystical images of imperial qualities – that other artists and connoisseurs used (Evans 1973; Kaufmann 1988; Debus 2001). Croll and de Boodt exemplify a spectrum of chymical physicians associated with the court who found inspiration in Rudolf's massive *Kunstkammer*.

At one end of the spectrum, Croll represents Paracelsian physicians associated with the court, but not formally employed by it. Significantly, Croll gained the court's approval and thus the weight of court legitimacy for his *Royal Chymistry*, even securing "samples of medicinal earths" from the emperor's collections for his chymical pharmacopoeia that comprised the second part of his tripartite book (Moran 1991c; Findlen 1997: 214; Fucíková 1997). Prior to this set of recipes, his text provided a pansophical examination of the Creation, as explained by microcosm–macrocosm connections, the religious physician's grace-driven obligation to observe and interpret nature, and the view that medical diseases were related to *semina* and *astrum*, views he derived from Paracelsus, "Basil Valentine," and Petrus Severinus (Debus 2001; Shackelford 2004; Moran 2007). These two parts of his text were followed by a treatise on signatures that explained how to uncover the hidden virtues of divine signatures inherent in natural things for medicinal use by performing chymical operations with fire (Debus 2001; Moran 2007). Croll portrays one type of Paracelsian

FIGURE 7.2 Collaborative efforts in the Hapsburg orbit: title page of Oswald Croll, *Basilica chymica* (1609), illustrated by the court artist Aegidius Sadeler, edited and used by Johannes Hartmann for the latter's *laboratorium* course. Wellcome Collection, London CC BY.

physician engaging in empirical and pansophical topics, belonging to humanist circles associated with the court, and gaining precious authority for his chymical views through court privilege rather than imperial patronage.

At the other end of the spectrum is the formally employed *Leibärzte* de Boodt, who, although chymically inclined, rejected aspects of Paracelsianism. Published in the same year as Croll's text, de Boodt's *Gemmarum et lapidum historia* (*Description of Gems and Stones*) classified over 600 minerals in Rudolf's collections and complemented these with their inherent divine or malevolent virtues. Moreover, he developed an unprecedented "three-color primary system as a basis for color mixing" based on gem color analysis (Parkhurst 1971: 4). Extrapolating upon the concept of an "architechtonic spirit" related to seminal theories, he explained growths of minerals and fecundities of vegetables during the Genesis Creation (Hirai 2007). De Boodt rejected the concept that metals were reducible to prime matter, yet he retained ideas of microcosm and macrocosm (Purš 2004).

Within this spectrum are physicians like Tádeas Hájek (Hagecius), who had served previous Habsburg emperors and remained influential within aristocratic and courtly circles. As an astronomer, astrologer, physician, and chymist, Hájek embodied the Hermetic trope "by looking up, I see down": by "looking up" to record celestial events, measure astronomical influences, and cast prognostications of celestial portents, he "saw down" to witness the spagyric arts, iatrochemists, and occultists whom he hosted, like John Dee and Edward Kelley, who promised to reveal novelties in both realms (Evans 1973; Fucíková 1997; Szönyi 1997). Chymists like Dee, Kelley, Johann Müller, and Michael Maier arrived at court hoping for imperial patronage and received an audience or title from the emperor, but few obtained actual imperial contracts (Prinke and Zuber 2018).

Historians distinguish chrysopoeian (gold-making) chymists like Sendivogius, who retained a contract to provide recipes for transmutation to the emperor, from *Leibärzte* like de Boodt, who were provided salaries and pensions (Nummedal 2007; Prinke and Zuber 2018). Sendivogius represents the influential "nonmetallic school of chymistry," which sought the solvent for making the philosophers' stone in rain and dew rather than in Philosophical Mercury, as the "Mercurialist school" proposed (Principe 1998). Expanding the Hermetic "as above, so below" trope to include dual suns, Sendivogius' *Novum lumen chymicum* (1604) analogized the "terrestrial sun" to a great distillery that condensed earth's elements into a sal-nitrum or universal salt, which, together with an implanted universal seed, was "sublimed to the surface" as a vapor by the "celestial sunrays" and then carried away into the dew and rain (Debus 1977; Prinke 1999; Hirai 2007). Promising to demonstrate how to make the philosophers' stone in exchange for financial gain (Rudolf's promise of carved mineral stones), Sendivogius managed to postpone the chymical demonstration for a decade before departing with his secrets and well-being intact (Prinke and Zuber 2018).

Physicians, *Leibärzte*, and chymists were valuable at Prague as court liaisons. Well-connected to the Bohemian chymical circles of wealthy aristocrats like the Rozmberks, Croll served as an agent of his former employer Christian of Anhalt to Rudolf's court (Evans 1973). Ancel and Bernaud intervened between Paris and Prague (Evans 1979); Sendivogius connected Paracelsian circles in Krakow and Prague and served as a "diplomatic envoy" between Rudolf and the Polish court of King Sigismund III Vasa (Prinke and Zuber 2018: 328). And Hajek's epistolary exchanges persuaded his correspondent at Uraniborg, the astronomer Tycho Brahe, to move to the Rudolfine court when Tycho's relations with the Danish court soured (Mosley 2007).

Material technologies: court laboratories and mining concerns

Beyond the quests that drove spiritual, artistic, political, and pharmaceutical energies described above, emperors and princes in the German territories made substantial improvements in the material technologies of laboratories and mining in their efforts to alleviate costly war debts and strengthen emerging state commerce. Widely acknowledged throughout Europe, the German mining industry had a long history of assaying and smelting difficult-to-process ores. The most scholarly treatise on mining and metallurgy during this period, the 1556 *De re metallica* (*On Metal*) by the physician Georg Agricola, provided information on processes, equipment, and techniques; the book was translated into several languages and acquired great international renown. German miners gained prestige as innovators and advisors on mining, metals, and minerals. Queen Elizabeth I, excited over prospects of extracting copper from English mines, hired German miners to transmit technical expertise to English mining managers, a promising collaborative project that failed due to managers' inability to trust foreign advisors, coupled with miners' incapacity to translate technical expertise to novices (Ash 2004).

Princes concerned about mining invested in chymical knowledge as a means to improve the extraction, smelting, and separation of metals and the perfections of metals into higher species. Entrepreneurial alchemy at the courts of Dukes Friedrich I of Württemberg and Julius of Braunschweig-Wolfenbüttel exemplifies the confluence of technical expertise, economic productivity, and chymical knowledge expected of, and promised by adepts: "The close connection between the separation and transmutation of metals was the key to metallurgical and alchemical work, and it marked the space in which mining entrepreneurs found alchemists to be of the greatest use" (Nummedal 2007: 90). Princes signed precise contractual agreements with chymical laborers to enhance economic growth in territorial mining investments; Duke Julius signed a contract promising monetary reward, a laboratory, plus lodging (and a horse) in exchange for transmutation activities to maximize the profitability of his interests in copper and silver mining concerns in the Harz Mountains. Dukes Julius and Friedrich prosecuted a few of these same court alchemists on charges of fraud, underscoring the gravity attached to matters of expertise. Despite the terrifying executions that ensued, in which some alchemists were drawn, quartered, and burned at iron gallows, they continued to obtain contracts with princes who paid for advice.

Chymical cultures in court laboratories grew in concert with distributions of labor within specific physical settings. Courtly laboratories tended to be more like artisanal workspaces than today's scientific institutional settings: they could be in cellars (as, for example, at Tycho Brahe's Uraniborg), in a hospital, at a noble's distant estate, or in a separate location, such as the Old Summer House in the ducal gardens at Stuttgart that Duke Friedrich of Würtemberg had converted into a laboratory (Hannaway 1975; Nummedal 2007; Klein 2008). Medicinal, metallurgical, and transmutational labors were dispersed among a variety of chymical workers, including pastors, doctors, barbers, managers, tinkers, porters, scribes, chamberboys, bookkeepers, scholarly chymical counselors, and inspectors (Moran 1991c; Nummedal 2007). Salaries, contracts, and employment agreements display a hierarchy of authority over specialized work staff in which well-paid alchemists provided recipes to laboratory workers, who completed assigned procedures for a nominal salary plus food and clothing (Nummedal 2007).

Different types of labor were distinguishable by where they were distributed throughout courtly spaces. Tara Nummedal's reconstruction of a sixteenth-century laboratory in the Old Summer House's second-floor, windowed workspace reveals the types of material labor performed there: fireplaces, portable copper furnaces, and large and small copper ovens indicate activities including the melting, smelting, and distilling of metals and materials. In addition to two workspaces, an attic space for chymical supplies and equipment held assaying balances, copper baths, ceramic vessels for metallic separations, and glass vessels used for various distillations and separations. Laboratory assistants worked in places with oversight, such as castle kitchens or the Old Summer House, whereas chymical advisors conducted work in less supervised laboratories, such as the nearby New Hospital. Famous chymists like Sendivogius, while under contract with Friedrich, worked independently at an estate outside of Stuttgart (Evans 1973; Nummedal 2007).

The interest in chymistry as a means to maximize material resources, especially its potential to increase mining and metallurgical profits, continued throughout the seventeenth century with emperors and princes hiring chymical advisors for commercial purposes. The Habsburg chymist Johann Joachim Becher followed decades of university study on chymical theory with laboratory work as a means to investigate the "generative principle" noteworthy for the transmutation of base metals into gold and to learn methods to "mature, improve and transmute" metals (Smith 1994: 202). Thus, courts were conducive to any number of chymical cultures: displays of power through glassworks, pottery, and sculpture, the spiritual and political engagement in chymical philosophy, the scholarly and practical economies surrounding transmutation, and/or the commercial risks involved in exchanges of mining and metallurgical productions.

Given what we understand today regarding the impossibility of metallic transmutation, one may wonder why princes continued to hire chymists if they were unable to perform this feat, or whether all chymical practitioners were imposters or con artists. Recent research by anthropologists, historians, and modern chemists provides some insight into questions about historical actors' claims. Practical information about metals, minerals, and chemical reactions can be reconstructed by carefully investigating historical inventories, recipes, chymical waste, archaeological remains, and architectural illustrations (Soukup 2007). For example, X-ray diffraction demonstrates that certain medicinals were produced at the famous alchemical laboratory in Kirchberg am Wagram, Lower Austria, as well as separations of gold and silver by use of nitric acid (Soukup 2007). Anthropologists analyzing crucible remains have identified specific procedures; for example, residues of magnetite crystals and alkali oxides within a silicate layer suggest operations that aimed at searching pyrites for gold. Evaluations of crucibles from Hesse (Prince Moritz's homeland) determined that the reputed superiority of these widespread, coveted vessels resulted from "an aluminum silicate that was not formally discovered and named until the twentieth century," which resisted corrosion in strong acids and survived firing at extremely hot temperatures (Martinón-Torres 2007: 158).

As for chymists' belief in transmutation, Lawrence Principe has raised considerable awareness of early chymists' skills, perseverance, and earnestness by means of his successful reproduction of several steps of an antimony-containing medicinal recipe needed to produce the elusive philosophers' stone using "Basil Valentine's" 1604 *Der Triumph-Wagen antimonii* (*The Triumphal Chariot of Antimony*; Principe 2013). Principe's reworking of this procedure, while not endorsing transmutation, revealed a colorful procedure nonetheless, namely the volatilization of gold chloride, a delicate operation that would challenge today's chemists with modern equipment (Principe 2013). Such studies indicate that some chymical adepts were conducting practical trials and producing provocative results, especially to those in positions of support, such as prince–practitioners like Moritz and his circle, the Habsburg emperors, and the German dukes, all of whom were excited by prospects demonstrable within chymical laboratories. These courtiers and physicians, who conducted and deciphered investigations for theoretical, spiritual, medical, and economic uses, soon expanded their scope to bridge workspaces at courts with those at universities.

CHYMISTRY AT EUROPEAN UNIVERSITIES

In the late sixteenth and throughout the seventeenth centuries, princes, prince–practitioners, and religious reformers expanded chymical cultures beyond courts by using universities to promote educational restructuring, transform botanical apothecary shops into chymical distilleries, and formalize chymical instruction through the rigors of scholarly study and disputation. These historical actors' interests were by no means similar; distinct university settings produced distinct scholarly traditions. Yet three aspects united chymical cultures: the role of chymical physicians as liaisons between courts and universities, the expansion of chymical texts into university curricula, and the introduction into the university of a dedicated space for conducting practical work. This dedicated space was an amalgamation of the courts' artisanal, metallurgical, and transmutational workspaces, which transformed variously into early academic laboratories (Klein 2008).

Chymistry became part of medicine at universities where pedagogical reformers subjected chymical substances to theoretical scrutiny and explicated ancient and contemporary chymical texts. This interplay of physicians, laboratories, and texts occurred in different sociopolitical contexts. In Marburg and Spain, statecrafting of curricula occurred by princely or royal decree; in Wittenberg, chymistry was shaped by reformed curricula; in Jena and Leiden, chymical reform emerged from critiques of Paracelsian and Galenic medicine; and in Prague, chymistry emerged out of negotiated Catholic–Lutheran curricular reform. As iatrochemistry spread to other universities, they became nodes within numerous crosscutting networks: teacher–student communities who transmitted laboratory and medical *experimenta* and chymical physicians who transitioned within court–university networks. While other developments could be cited, these few examples evince the difficult and contingent process of discipline formation in early modern universities.

The University of Marburg

The powerful effect of princely patrons upon curricular reform is most clearly exemplified by Prince Moritz of Hesse, who used his territorial governance over the university as an effective tool of statecraft. In his efforts to build an alliance of Calvinist and Lutheran princes against Catholics, Moritz established a university founded upon the scholarship of Hermetic, Neoplatonic, and crypto-Calvinist worldviews, creating an "intellectual pattern of the Kassel court" by appointing several self-selected chymically inclined physicians and philosophers to the University of Marburg faculty (Moran 1991c: 35–67). To the chair of Natural Philosophy Moritz elected an occult philosopher (Rudolf Goclenius), and as professor of Philosophy and Theology he assigned Raphael Eglinus, who defended the scriptural basis of alchemy as a balsam for the widely prophesied eschatological end-times. However, his promotion of Johannes Hartmann in 1609 to teach the theory and practice of chymical medicine ushered in the new discipline of iatrochemistry (Moran 1991a).

Hartmann fundamentally altered the traditional university medical curricula at Marburg through lectures that combined traditional Galenic medicine with Hermetic–Paracelsian philosophy, and more strikingly he supplemented these with semesters spent in a novel university space, a *laboratorium*, structured for a professor to provide instruction in hands-on *spagyric* chymico-medicinal recipe preparations (Moran 1991a). *Spagyria*, a process that involved breaking substances down (analysis) into their component parts and recombining them into new products (synthesis), went beyond merely procedural steps and embraced key concepts within Paracelsian theory involving medicine, metallurgy, artisanal crafts, and even the *ex nihilo* moment of Creation (Smith 1994; Newman 2006a). Once recovered, component parts offered exciting prospects for medicaments and transmutation.

King Philip II of Spain also introduced iatrochemical medicament preparation into the University of Valencia's medical curricula by promoting the work of chymical distillers. The court physician and experienced distiller Lorenzo Cózar created a position at the university for a chair of secret remedies (Moran 1996; Rey Bueno 2015). Despite superficially appearing to be an extension of a pharmaceutical preparation, chymical distillation provoked controversy; medicinal distillation was carried out by empirics, charlatans, court jesters, women confectioners, and brewers of aquavit, areas often governed, but distained, by university-trained physicians (Eamon 1994: Gentilcore 1995). And when advice to include iatrochemistry within curricula was offered by licensed physicians in Italy, England, and Germany, it mostly went unheeded (Cook 1986; Eamon 1994). Thus Philip's support for distillers helped raise distillers' status, leading to a period of "greater visibility of chymical practice in early modern Spain" that reemerged during the late seventeenth-century reign of Charles II (Rey Bueno 2015: 135). Similarly, it was due to the powerful patronage of the prince–practitioner Moritz that Marburg's laboratory included distillation.

Bruce Moran's study of Hartmann's 1615–1616 laboratory diary provides a sense of an early modern iatrochemical classroom: leaving their swords and insolence at the door, students entered a workspace containing distillation equipment, a balance, and various "glass and earthen vessels ... water and sand baths and ovens ... [and] a variety of chemical substances". Opening the textbook – Croll's *Royal Chymistry* – Hartmann might guide them in calcinating vitriol, distilling urine, purifying opium tinctures, or preparing antimony, while directing them to take good notes: which retorts, alembics, or chemicals were used, what "temperature and lengths of time" were materials fired, and in which type of ovens (Moran 1991c: 62–64). The new *chymistry* that joined hands-on practice with the spiritual, Hermetic, and vitalistic mindset of Moritz's circle embodied a distant cousin from later *chemistry*, yet one with a vague family resemblance.

The early modern *laboratorium* was obviously very different from today's university laboratory; nonetheless, its entry into universities as a pedagogical space marked a new development in chymical sciences. Ursula Klein has drawn attention to the transitory nature of this space, as a courtly kitchen, or within a professor's house, or as a revised workspace. Part apothecary, part artisanal workshop, it constituted the place of manual labor, often involving the help of a laboratory assistant, where heavy furnaces could be fired, ample wood and water easily acquired, and chemicals, glass, and earthenware vessels safely stored (Klein 2008). Since chymical procedures entailed sooty, stinking, caustic, and perilous operations, these had to occur in a dedicated and secure space (Smith 1994; Klein 2008). As a key turning point for epistemology, the advent of chymical laboratories in university instruction created opportunities to challenge fundamental aspects of natural philosophy.

The University of Wittenberg

Not only princes, but also religious figures shaped political ends with pedagogical tools, including leaders at the highest levels of hierarchy at the University of Wittenberg, the flagship of Lutheran education. Not only did Paul Luther practice chymical medicine, but his prominent father, Martin, spoke of it favorably (Nummedal 2013; Klein 2016a). Caspar Peucer, professor of medicine, university rector, and son-in-law of the university's cofounder Philip Melanchthon, collected a sizable chymical library; and Tycho Brahe, who spent sufficient time as a student there to lose part of his nose in a duel, brought his chymical and astronomical interests to Wittenberg. Returning in December 1598, following his tragic fall from the heights of patronage that funded his Hven castle, Uraniborg, but prior to his recovery and spectacular ascent to the Rudolfine court as Imperial Mathematician in 1599, Tycho resided and erected a chymical laboratory at the home of a professor of medicine, Jan Jessenius, who resided at Melanchthon's and Peucer's former home (Christianson 2000; Mosley 2007). Even Wittenberg's famous mathematics professor and disciple of Copernicus, Georg Joachim Rheticus, favored chymical medicine and practiced it at Krakow (Klein 2016b: 295). To whatever extent these examples demonstrate a culture in which chymistry was accepted by the most prestigious minds of the university, actual instruction in the theory and practice of chymical medicine began with Daniel Sennert, who matriculated as a student prior to his professorship there.

Initially a traditional Aristotelian, Sennert's experiences as a chymical laboratory instructor caused him to undergo a striking transition from his antichymical stance to a defender of chymistry and atomism by 1619

(Newman 2006a). Sennert's medical teaching merged aspects of Galenism and Paracelsianism; he neither adopted the *semina* concept of disease that marked Paracelsian medical theory, nor did he qualify as Galenist, since instead he adopted the *tria prima* (Debus 1977). Sennert began offering a *collegium chymicum* – a hands-on laboratory course, perhaps at his home – in 1616 (Newman 2006a; Klein 2016a). Unlike at Marburg, where Hartmann's laboratory notes survive, the actual coursework is unknown; however, illustrations accompanying Sennert's 1611 *Epitome* and his correspondence suggest he engaged students in various experimental projects, such as the production of antimonial and mercurial medicaments (Klein 2016b). Letters exchanged between 1619 and 1637 depict a vibrant laboratory community in which students and teachers collectively shared trials over furnaces, cucurbits, and earthenware; this textual evidence was recently verified by archaeological digs (Klein 2016b). Sennert's courses integrated manual and intellectual labor, thereby creating new avenues to critique theoretical premises in light of experiential laboratory results.

The discovery of powerful acids like nitric acid, which could dissolve metals, utterly changed the potential for demonstrating and exploiting the analysis and synthesis of metallic components and had profound consequences, since they threatened to erode traditional scholastic matter theory as it was taught in the schools using medieval interpretations of Aristotle by Thomas Aquinas. Sennert's publications offer insight into the ways in which chymical laboratories challenged students' understanding of epistemic claims (Newman 2006a; Klein 2016a). In 1619, Sennert experimentally demonstrated the existence of atoms by means of a procedure called "reduction to the pristine state," an experiment in which silver was dissolved in nitric acid to form a blue liquid and then was precipitated out with salt of tartar to form silver again (Newman 2006a: 112). A key principle of Aristotelian matter theory, underpinning claims about the structural composition of natural or artificial things, held that for a thing like silver to exist it must be composed of a unique and unitary *substantial form* and *elementary matter*. Yet this "reduction" experiment contradicted what conventional matter–form theory predicted. Under scholastic matter theory, the nitric acid that dissolved the metal should have completely destroyed silver's *substantial form* and created a new form – a true mixture of silver and acid – exhibited by the appearance of the blue liquid, the consequence of the metal's dissolution (Newman 2006a: 99–100). At this point Thomistic Aristotelians would perceive that it was no longer silver but a new mixture. Further troubling this prediction was the reemergence of silver from the solution once a precipitating agent (salt of tartar) had been added, since silver's reappearance suggested that it had not been destroyed at all, but rather had continued to exist as silver throughout the procedure. The reappearance of silver cast doubt upon the belief that the new form of homogeneous blue liquid had destroyed the initial metallic form of silver, its characteristic qualities, and the matter into which qualities of silver inhered. Rather than dismiss Aristotelian matter–form theory, Sennert devised a compromise explanation: to the senses, the acid and metals had broken into such tiny bits that they were imperceptible, yet they retained their form and matter when they cohered in liquid (Newman 2006a). This concept of semipermanent forms in tiny atoms, although not original to Sennert, exemplifies the productive intersection of a chymical practice and experimental theory that would intrigue and occupy a later chymist, Robert Boyle, a half-century later (Newman and Principe 2002).

The University of Jena

The drive to create a form of chymical medicine that would be acceptable as a discipline within the faculties of arts took root most favorably at Jena, where teaching medical philosophy together with laboratory practice produced several professors of chymical medicine. Aristotelian, Galenic, and Paracelsian scholars alike played roles in shaping chymistry into a discipline by not only emphasizing precision in language and reconceptualizing traditional theories of form and matter, but also by supporting chymical practices with theoretical axioms. None were more exemplary of this than the outspoken polemicist, chymist, physician, and author of one of the earliest chymical textbooks, *Alchemia* (1597), Andreas Libavius. Adamantly rejecting Paracelsian medicine, Libavius advocated instead for an Aristotelian-based chymical medicine for universities, which included the predominant method of scholastic disputation (Hannaway 1975; Moran 2007). Trained at the Lutheran universities of Jena and Wittenberg, Libavius grounded his chymical views in Galenic–Hippocratic medicine, yet he included aspects of Hermetic, magical, and occult worldviews. Annoyed by the Paracelsians' reliance upon divinely based revelatory wisdom as a means to legitimize chymistry, Libavius resolved to show that validation rested solely with universities, even to the exclusion of princes and emperors, and in this crusade he circulated his contentious views, hurling an antagonistic arsenal of letters and *ad hominem* attacks at those with whom he disagreed and agreed (Moran 2007).

To counter the Paracelsian attack on reason-based epistemology, Libavius set forth specific ways that chymical medicine could become a scholarly based discipline within university faculties of arts, where students would merge philosophy and laboratory crafts to become scholar–artisans (Moran 2007: 42). Significantly, he saw chymistry as standing on its own, as a discipline with its own principles, detached from medicine and pharmacy (Moran 2015). Writing to the Dean of Medicine Zacharius Brendel Senior at his *alma mater*, Libavius recommended such practices as submitting to public debate any results either witnessed of others' manual craft labor or performed with one's own hands at one's laboratory (Moran 2007; Moran 2015). Such recommendations, so closely representative of the new experimental philosophies of the *moderns*, it is worth noting, were inscribed by the pen of an Aristotelian–Galenic *ancient* apologist (Clucas 2007; Moran 2015).

Although it is unknown whether Brendel was influenced by Libavius, he shared his fellow alumnus's goal of shaping "the art of chymistry ... while preserving the principles of physics and medicine," making it "suitable for our [academic] studies" (Moran 2007: 37). Brendel was first to establish laboratory instruction at the University of Jena in 1612, inaugurating a course in chymical lectures and laboratory instruction for medical students that his son and namesake continued. Zacharias Brendel Junior and other professors at Jena (among other universities) dealt with the dearth of chymical textbooks by writing their own; Brendel's 1630 text, *Chimia in artis formam redacta* (*Chymia, Having Been Rendered into a Form of Art*), provided instruction on various types of remedies, chymical fires, and operations, in addition to an assessment of the reliability of certain authors, like "Basil Valentine" (Partington 1961). Yet it was the Brendels' successor, the professor of anatomy, surgery, and botany Werner Rolfinck, who, as the first Chair of Chymical Exercises (in 1637), created a more effective method to legitimate chymistry within the university (Debus 2001).

It was Rolfinck who most fully molded iatrochemistry into a university discipline at Jena, complete with a scholarly disputational format, a formalized use of chymical textbooks, and a scholastic positioning of it. However, for Rolfinck, chymistry was not an independent discipline; he viewed it as subalternate to medicine, since chymistry was consistent with medicine's goals. Rolfinck envisioned a compromise between (to him) acceptable aspects of Paracelsianism, including the *tria prima* dogma, and (to him) objectionable aspects, such as the Hermetic doctrine of the macrocosm–microcosm and Paracelsian notions of homunculus, palingenesis, and metallic transmutation (Partington 1961). Terming these latter "chymical non-entities," Rolfinck's criticism drew ire from later chymists Robert Boyle and Johann Joachim Becher (Principe 1998). Yet in his staunch defense of iatrochemistry against Galenists, Rolfinck was one of the first to establish a text that grounded theoretical arguments about chymical laboratory operations, offering "academic arguments in the form of dissertations" so that students could apply both theory and practice in a disputational academic format (Powers 2012). His student Georg Wolfgang Wedel succeeded him as chair and followed these efforts to improve chymistry by clarifying symbols, standardizing language, and establishing rules for chymistry (Garber 2015).

The Carolinum University, Prague

Chymistry in Central Europe faced considerable difficulties in acceptance into the curriculum, especially in overcoming neo-Aristotelian arguments about the basic composition of natural things. Efforts toward curricular expansion to include iatrochemistry and chymical philosophy at Prague's Carolinum

University were abruptly thwarted by its faculty's involvement in the early events that instigated the Thirty Years' War (1618–1648). The religiously tolerant and intellectually open climate of the Rudolfine court that had warmed to iatrochemistry and Neoplatonic, Hermetic, and chymical philosophy at both the court and the Carolinum cooled considerably in the destabilizing aftermath of the Protestant defeat at the Battle of White Mountain in 1620. The institution that had served the interests of both Protestant and Catholic students of the Bohemian bourgeoisie with a Melanchthonian, neo-Aristotelian Faculty of Arts curriculum since the 1580s became entangled in the early uprisings when some faculty supported the Calvinist Palatinate Elector Frederick V's ascent to the throne of Bohemia (Evans 1979; Pesek 1991). After the Catholic international army defeated the rebels, radical faculty associated with the insurrection were assassinated, including the "spokesman for the Bohemian revolt," the public anatomist and medical instructor Jessenius (Evans 1979: 108). Consequently, the university became an incendiary target for the new emperor, Ferdinand II, who had to be "narrowly persuaded not to raze the rebel Carolinum to the ground and build a home there for the public executioner" (Evans 1979: 224). To assure conformity, political and theological, the Carolinum was restructured, its faculty became Catholic, study abroad was curbed, and its autonomy and administrative oversight were ceded to the Jesuit Collegium Clementinum, actions negatively impacting the curricular place of chymistry (Evans 1979; Cornejová 1997).

In the prewar years, Neoplatonic and Hermetic training was carried out by instructors like Jessenius, and iatrochemical training was offered through apprenticeship, established by the former rector and Paracelsian physician Martin Bachachius (Cornejová 1997). This was standard practice. Most skills in chymistry that medical students acquired throughout the German territories were offered privately; in Helmstedt, Duncan Burnett and Gregor Horst offered courses in private houses, as did Henning Arnisaeus in Giessen (Klein 2016b). At Prague, Jan Marcus Marci, who matriculated at the Carolinum in 1618, described his private chymical training as an apprenticeship with the alchemist Georgius Barschius that continued after the restructuring (Garber 2005). The new Jesuit rector, Martin Santinus, who was initially skeptical of chymistry, even came to "love *chymia*" and find value in iatrochemistry (Marci 1662; Garber 2005). Far more difficult was navigating a space for interpretive creativity when the Carolinum's medico-chymical courses conflicted with Clementinum's Faculty of Arts courses in philosophy.

Similarly to Severinus, Sennert, Libavius, and other chymical physicians who struggled to find a compromise between Aristotelian hylomorphism and chymical philosophy, Marci interpreted Aristotelian prime matter (based on Duns Scotus) to exist as independent material corpuscles, but he denied the concept of substantial forms in favor of "seminal ideas" derived from a broad set of ancient and contemporary authors (Mocchi 1990). Merging Neoplatonic philosophy, Hermetic microcosmic–macrocosmic connections, and Paracelsian *semina*, Marci proposed that seminal ideas, or seeds, directed the generation of all natural things and transmitted celestial properties to terrestrial beings through light (Garber 2005; Garber 2007). Marci's rejection of substantial forms in favor of seminal ideas and his Scotist interpretation of prime matter were considered heterodox by Clementinum's Dean of Theology, Roderigo Arriaga S.J. The disagreement between Arriaga and Marci would have constituted a mere scholarly disagreement had it not been for the fact that Marci was promoted to Carolinum's Dean of Medicine (1638–1665), and the two deans represented opposing sides during the decades-long negotiations over the two universities' merger into the Charles-Ferdinand University (1654). In this undertaking, Marci served as liaison between both court and university and papal and imperial courts (Cornejová 1997). Fortunately for the medical faculty (and, consequently, iatrochemistry at Prague), Marci succeeded in securing his faculty's autonomy from the Faculty of Arts in the merger arbitrations, despite Arriaga's disapproval. As such, Marci exemplifies *Leibärzte* who served as liaisons by bridging courts, universities, and artisanal workshops.

Marci was well connected to aristocrats with chymical interests and to Prague's imperial workshops through his in-laws, the Miseroni, one of Hradcany's most prosperous artisanal families of glass cutters, gem engravers, stone sculptors, and later trusted curators of Prague's post-Rudolfine *Kunstkammer* (Fucíková 1997). Perhaps due to his chymical interests and his wife's connections, Marci rose to *Leibärzte* to Holy Roman Emperor Ferdinand III when in residence. Later, he joined Emperor Leopold I's courtly tours of the German territories, in which as *Leibärzte* Marci advised on mineral baths and medico-chymical matters (Fucíková 1997; Garber 2005). The Habsburg prince–practitioners were crucial to chymistry's expansion in Central Europe; they patronized chymically inclined *Leibärzte*, hired chymical advisors, and attested to having witnessed transmutations (Smith 1994). Unlike his father, during the merger of his namesake university Ferdinand III chose not to intervene in administrative matters; thus, chymical medicine entered this university's curricula more through mediation than patronage.

Chymical learning in Paris, Oxford, and Cambridge

In England and in France, with the exception of the University of Montpellier, chymistry did not become part of the university curriculum until late in the seventeenth century. Instead, France represents a culture of chymical learning that spread through the didactic tradition, a means of creating a teachable subject, by use of textbooks created for specific courses (Hannaway 1975; Clericuzio 2006). This didactic tradition in Paris emerged in two types of extracurricular courses: those created by individual apothecaries and conducted privately at their shops or homes and those offered by individuals receiving support from royal physicians and performed publicly at the *Jardin du Roi* (Clericuzio 2006). Textbooks that accompanied such courses proliferated through students' course notes that in turn became textbooks. Continual pirating of texts also contributed to broad circulation; the Huguenot apothecary Jean Beguin's textbook on pharmaceuticals is a case in point (Johns 1998). Following Libavius' advice (as well as lifting entire verbatim passages from him), Beguin published his lectures in 1610 as *Tyrocinium chymicum* (*Apprenticeship in Chymistry*), which was reprinted numerous times with later editions growing three times in size (Debus 1991; Clericuzio 2006; Powers 2012). Following Beguin's lead, French teachers of chymistry in the 1630s and 1640s printed texts based mostly upon collections of medical recipes and assaying techniques, often from students' lecture notes that circulated widely and helped spread interest and technical knowledge of chymical practices (Shackelford 2004; Clericuzio 2006). By the 1640s, several chymists were teaching public courses at the *Jardin du Roi* laboratory, some of whom displayed Severinian philosophy – by 1650, Flemish chymist Jan Baptiste van Helmont's views were so acceptable that half of the medical faculty accepted iatrochemistry (Clericuzio 2006). This textbook tradition, which supplemented public or private courses, impacted the growth of chymistry by making it available to persons beyond university walls or courtly artisanal workshops and to nonchymists: "By the middle of the seventeenth century, the audience for didactic chemistry had expanded to include physicians, natural philosophers, writers, artists, mechanics, and gentlemen savants" (Powers 2012: 43).

In England, where a rich culture of private chymists, apothecaries, gentlemen savant societies, and informal public chymistry had emerged in the mid-seventeenth century, it was not until the 1680s that chymistry first entered the university at Oxford through a bequest from the chymist Elias Ashmole, who funded a chymical laboratory, dual positions for a museum curator, and an "Ashmolean Professor" of Chymistry (Guerrini 1994). At Cambridge, chymistry officially entered university curricula in 1702, although since the early 1680s chymical operations, metallurgy, and pharmacopoeia had been unofficially taught at an outdoor laboratory to apothecaries and physicians (Guerrini 1994: 188; Knox 2005).

The University of Leiden

As the examples from the universities of Wittenberg, Jena, and Prague indicate, chymistry had to be subjected to the philosophical rigor of traditional scholastic matter theory to gain entrance into university medical curricula. In turn, scholastic matter theory began to be subjected to chymical experimentation and chymical theories about matter, which transformed or even eliminated traditional scholastic theories about form and matter. Further efforts to shape chymistry into an academic discipline occurred at Leiden, where fashioning chymistry as an accessory to medical education resulted in curricula in which chymistry became the underpinning for medicine, especially as chymical analyses could explicate physiological processes. Understanding biology by means of chymical experimentation had been undertaken by earlier authors, such as van Helmont and Johann Glauber; however, the professors of medicine at Leiden established chymistry as a discipline that steered discoveries in anatomy and physiology onto new shores. Methodological innovations in chymical trials created new ways to dispute and overturn prior chymical and natural philosophies.

At the University of Leiden, iatrochemistry emerged out of Franciscus Sylvius de le Boë's successful efforts to persuade authorities to build a chymical laboratory at the university, which he used as a site for tutoring interested students and as a base for chymical investigations (Powers 2012). Sylvius altered Leiden's medical curriculum by complementing it with chymical operations like fermentation and neutralization as a means to explain the physiological processes of hunger, thirst, digestion, and fevers; furthermore, he connected chymistry to anatomical findings in the lungs, heart, and brain (Debus 2001). In Sylvius' iatrochemistry, acids and alkalis within bodily fluids featured prominently as the crucial determinants of bodily function, a concept he derived from van Helmont and Glauber (Partington 1961).

While an extended discussion of van Helmont and Glauber (who were not affiliated with courts or universities) is beyond the scope of this chapter, a few examples provide context for Sylvius' activities at Leiden. Van Helmont was an early seventeenth-century Paracelsian adept, who accepted Severinus' concept of *semina* and *archei* as the basis for bodily organs and disease and aspects of the *tria prima*, yet viewed the latter as either proximate ingredients or as products of chymical operations rather than *a priori* principles, and he dismissed the microcosm–macrocosm analogies (Debus 1997; Moran 2005; Principe 2013). Van Helmont was especially important for innovating experimental methodologies, such as the use of mass balance: assuring that the weight of substances entering into a chymical operation equaled the weight of products recovered, a concept that presumed the conservation of mass (Newman and Principe 2002). In terms of the chymistry of the body, van Helmont proposed that digestion resulted from stomach acids, which could be neutralized by alkalis (Moran 1996). In the mid-seventeenth century, Sylvius adopted ideas from van Helmont and received assistance from Glauber. Glauber was a traveling autodidact who settled in Amsterdam and created a laboratory in which he experimented with acids and alkalis, invented new instruments, apparatus, and experiential methods, and wrote numerous treatises on what he observed therein (Smith 2004; Cook 2007). He taught briefly as an unsalaried lecturer at Utrecht; however, due to his renown as an innovative instrument-maker, apothecary, and chymist, many students apprenticed with him, including Carel de Maets and Jacob Le Mort, both of whom became students of Sylvius (Smith 2004; Powers 2012).

De Maets arrived in Leiden to become its new laboratory's first chymistry instructor for Leiden's medical students; however, his "appointment replicated the old model of chemical instruction at Leiden, but now the artisan chemist was simply brought in-house" (Powers 2012: 50). Rather than innovation, de Maets continued the prior tradition of training medical students on production and methods of making medicaments. De Maets and his pupil Le Mort, an apothecary and physician, had numerous students who created textbooks by publishing collections of their notes, which evince training of mostly recipes and chymical operations (Powers 2012). However, their pharmaceutical proficiencies and anatomists' skills enabled them to chymically analyze bodily fluids for evidence of acids and alkalis.

Further developing Sylvius' anatomical and chymical training, his students paradoxically laid the groundwork for overturning much of his acid–alkali theory of digestion. The skilled comparative anatomist Reinier de Graaf, who drew fluids directly from a living canine pancreas (through cruel expansion of instrument-assisted vivisection), submitted fluids to chymical and taste tests that uncovered inconsistencies with the acid–alkali prognoses (Ragland 2008). Expecting pancreatic juices to be acidic, de Graaf published his taste test *experimenta* vaguely confirming his mentor's prediction; yet later critics demonstrated that de Graaf's prediction was incorrect, since pancreatic juices were found to be slightly basic instead (Ragland 2008). These types of trials began to unravel many of the threads that spooled around Sylvius' iatrochemistry. Chief among these anatomists and chymical physicians, Herman Boerhaave advocated for chymistry to become foundational for medicine.

Boerhaave's chymical training developed over time through his extensive reading of chymical literature followed by chymical trials, which he used to root out errors in this literature rather than to master various pharmaceutical recipes and procedures (Powers 2012). Chief among his targets were the chymical philosophies of Sylvius and van Helmont; even his mastery of Helmontian analytical methodologies (like mass balance) was used to overturn chymical philosophies that he opposed (Powers 2012). More than previous skeptics, Boerhaave's fastidiousness in developing ever better analytical methods to accomplish these corrections led him to enhance his anatomy and natural philosophy courses through chemical practices adapted to include instrumentation (Powers 2012). Boerhaave's strategy would be imitated by future Enlightenment figures (Figure 7.3).

As chains of mentors and students developed chymical traditions at these universities, their students moved across Europe to impart chymical learning to new places. Sennert's student Rolfinck, for instance, initiated a long line of prominent students of chymistry at the University of Jena: Johannes Bohn (professor of anatomy and therapeutics at Leipzig) and Wedel (who succeeded his mentor Rolfinck as Director of Chymical Exercises). The latter taught Friedrich

FIGURE 7.3 Boerhaave giving a lecture at the University of Leiden. Title page, Hermann Boerhaave, *De comparando certo in physicis* (Leiden, 1715). Wellcome Collection, London CC BY.

Hoffmann and Georg Ernst Stahl, who founded chymistry at the University of Halle and influenced Enlightenment chymistry (Partington 1961; Debus 2001; Chang 2015). Marburg graduates branched out to other universities, developing textbook traditions: Daniel Becher to the University of Konigsberg; and Johann

Daniel Mylius to the University of Giessen, teaching iatrochemistry and writing chymical treatises (Klein 2016b). From Leiden, Boerhaave's students spread to the Viennese court and to universities in Göttingen and St Petersburg; and later medical students trained in Leiden transplanted Boerhaave's medico-chymical tradition to Edinburgh, Moscow, and Philadelphia (Powers 2012).

CONCLUSION

Between 1550 and 1700, chymical learning at courts and universities grew as texts, theories, materials, methods, and *experimenta* traveled through physicians from spaces of work to places of status. Late sixteenth-century courts at Marburg, Prague, Stuttgart, and Wolfenbüttel provided heterodox chymistry and iatrochemistry a place of legitimacy, where texts passed from libraries to hands, artisanal workshops expanded into chymical kitchens, and metallurgical and apothecary practices transferred among chymists, artisans, apothecaries, and chymical physicians. Transitioning between cultures at courts and universities, courtier–scholars translated chymical kitchens and apothecary shops into early laboratories. At Prague, Jena, Wittenberg, and Leiden, physicians exposed chymistry to the rigors of discipline formation. Scrutinizing scholastic philosophy by means of trials, chymists brought chymical and physiological *experimenta* to bear upon chymical theories. Laboratories became spaces where chymical results could challenge truth claims, whether of scholastic or of chymical origin. By the late seventeenth century, medical curricula at several universities included iatrochemistry, its status raised by the agency of liaison physician–professors joining universities or as *Leibärzte* uniting courts and universities. It is as if, by looking to courts, physicians saw new chymical methods, materials, and *experimenta*; by looking to universities, physicians saw a new chymical discipline enlisting methods, materials, and *experimenta*.

CHAPTER EIGHT

Art and Representation: *Skepticism and Curiosity for the Alchemist at Work*

ELISABETH BERRY DRAGO

Alchemy in the sixteenth and seventeenth centuries sat at a disciplinary nexus, a meeting point between philosophy, art, labor, and empirical inquiry. Its practices were embedded in countless economic enterprises, from mining and metallurgy to the making of dyes and pigments, to the production of chemical medicines, tinctures, and tonics. Alchemists also engaged in more experimental transformations and sought "truths" about the nature of matter and of human existence: what does it mean to create, to transform? To purify or perfect? Both alchemy's defenders and its critics struggled with these complexities and sought ways to sufficiently define alchemy, criticize it, or advocate for it. Responses to alchemical ideas ranged from delight to condemnation, curiosity to doubt. At times a single individual might express distrust for alchemical theory while embracing practical knowledge gained by chemical work. This was true of the German metallurgist Lazarus Ercker, who criticized the fool's pursuit of transmuting lead to gold, yet admitted that the "excellent, ancient, and useful" method of assaying metals originated from alchemy, which he described as an art of working fire (Ercker 1574: 9). In simple terms, alchemy's power to transform – and therefore potentially to falsify – made some people very nervous, while alchemy's ability to generate new materials, and to answer old questions, made others very excited. Sometimes, it did both at once.

This tension finds expression throughout the literature, theater, and art of early modern Europe, particularly in England, the northern and southern Netherlands, and the German-speaking territories. The fictional and pictorial subject of the alchemist at work began growing in popularity during the late sixteenth century and was ubiquitous by the early seventeenth century. Plays mocked the flowery allegorical language of false adepts and the greed of patrons, while laboratory scenes in genre paintings displayed fine details of broken glassware and dirtied crockery. Satires of alchemists, on stage or within the frame, have long been interpreted as evidence that the audiences they amused considered alchemy impractical, useless, or fraudulent. Scenes of workshop disarray (broken jars, dented pans, shattered glassware, and so on) have been viewed as unequivocal signs of alchemy's failures. Yet a careful examination of the pictures and texts as discussed in this chapter shows a plurality of approaches to alchemy. Some works offer sympathy to failed alchemists – others, harsh scorn. Some works celebrate alchemy as a bringer of harmony and benefit, while others decry a false alchemy hiding behind a veil of learned language. Recent historians of science have emphasized plural *alchemies* rather than a singular *alchemy*, as the goals and methods of its practitioners were vastly diverse (Linden 2003). Likewise, there was not one distinct way of picturing or writing about chemical work and chemical workers, but many.

It is little surprise that alchemy should have been so prolifically and diversely pictured during this period, an alchemical "Golden Age" in which chemical inquiry was woven throughout intellectual, economic, social, and political life. In late sixteenth-century Bohemia, the Holy Roman Emperor Rudolf II drew alchemists and natural philosophers to his court, shaping an international web of intellectual exchange and patronizing a generation of experimental empirical thinkers. These included Michael Maier, a disciple of the chemist–physician Paracelsus and an advocate for chemical medicine; the polymath Heinrich Khunrath, whose *Amphitheatrum sapientiae aeternae* (*Amphitheater of Eternal Wisdom*) contained fantastical illustrations of alchemical theories and idealized laboratory spaces; Polish physician Michael Sendivogius, who was believed to have successfully transmuted gold; Italian alchemist–theologian Giordano Bruno; English astrologer John Dee and his associate Edward Kelley; Czech alchemist Bavor Rodovský, who translated the foundational Hermetic text *The Emerald Tablet* into his native language; and countless others. While the alchemical pursuits of the court of Rudolf II were not coherent enough to constitute a "school" or academy, they helped set the stage for the next century's development of specialized scientific societies (Marshall 2009).

Among the most famous of these is the Royal Society of London for Improving Natural Knowledge, formed in 1660 and chartered in 1662 under the patronage of King Charles II. Passionately curious about alchemy, Charles II retained both a consulting alchemist – Nicaise LeFèvre, granted a suite of laboratories near Whitehall Palace – and a chemical physician, Thomas Williams. Yet the Royal Society did not emerge from the court: it was born of an informal network of scholars, empirics, and interested amateurs. These networks included researchers in Oxford and at Gresham College, London, the circle of the polymath Samuel Hartlib, and the connections of a young Robert Boyle (future associate of Sir Isaac Newton) and his sister, Lady Ranelagh, both of whom were involved in chemical experimentation (Principe and Newman 2002; Principe 2003).

Some members of the Royal Society – including Boyle – were deeply interested in alchemy, including transmutation or *chrysopoeia* (gold-making), although this fact would later often be denied and obscured. Some of these denials originated within the Society itself: its first official historian, Thomas Sprat, regularly derided seekers of the philosophers' stone (Principe 2013: 180–1). Likewise Joseph Granvill, an early fellow, condemned transmutation as a relic from a less enlightened age and stated that the Royal Society would refine chemistry from "dross" and at last render chemists "honest, sober, and intelligible" (Rattansi 2004: 364). But Boyle vigorously defended alchemy, corresponding with fellow alchemists (including the American-born George Starkey) and authoring the *Dialogue on Transmutation* around 1670. Within it, Boyle supported the viability of the philosophers' stone, claiming to have witnessed successful transmutation in the presence of a close friend, Edmund Dickinson, a royal physician. Several years later, in 1689, Boyle and Dickinson delivered impassioned testimony before Parliament regarding the legitimacy of transmutation, which led to the repeal of a 1404 English statue against gold-making.[1]

Artistic depictions of alchemy as either a beneficial or a suspect art were often determined by alchemy's relationship to gold-making. As the divide within the Royal Society makes plain, the quest for gold was an evocative (and polemical) subject, particularly during the transition period of the seventeenth century, as chemists across Europe sought legitimacy by distancing themselves from past practices. Gold-making received disproportionately high attention in the fine arts, a fact that is likely tied to such controversies. Where alchemy was shown to be characterized by gold-making, it was typically also characterized by greed, delusion, duplicity, and eventual poverty. Where alchemy was presented through other processes, such as distillation, assaying, metallurgy, or medicine, its representations were frequently more positive, or at least ambivalent. Distillers and assayers might appear as masters of busy workshops, while their counterparts in medicine and pharmacy were shown as scholarly experts diagnosing patients. Even more striking are paintings that visualize the relationship between alchemy and artisanal practice. This curious facet of alchemy's representation, I argue, is due to strong parallels between the chemical experimenter's quest to master nature and the artist's investigations of the same. These certain artists who produced images of alchemists as artisans appear to have been keenly conscious of alchemy's parallels and rivalries with art.

There is also the matter of alchemists' own use and creation of images and their deliberate constructions of idealized alchemical personas. Alchemists commissioned and produced images of equipment for practical purposes, but also pictured theories through symbolic diagrams – simultaneously revealing meaning to the initiated and concealing it from outsiders. Their writings also provided descriptions of "good" alchemists, praiseworthy and upright, in counterpoint to negative portrayals. The alchemist was one among many targets for satire: quack physicians, false merchants, and greedy tradesmen of all kinds appear throughout period imagery. But keeping a good reputation was an urgent matter. Consequences of faking *chrysopoeia* or breaking contracts with powerful clients could include fines, imprisonment, and death (Nummedal 2007). The sharpest condemnations of alchemical fraud were penned by alchemists themselves, who had the most to gain (or lose) in dispelling damaging myths.

ALCHEMISTS BOTH "FALSE" AND "TRUE" IN TEXT

The relationship between gold-making, fraud, and foolishness is evident in Sebastian Brant's famous satire, *The Ship of Fools*. First published in 1494 and republished in at least six editions across the next twenty years, the work was accompanied by woodcut illustrations, some carved by a young Albrecht Dürer. Brant's scathing account of "quite deceptive alchemy" is paired with a scene of alchemists wearing fool's caps, complete with bells and pointed ears. The alchemist is shown stirring a ladle in a basin over a furnace, but the scene is of trickery, not experiment: "Pure gold and silver doth it yield/But this in ladles was concealed" (Brant 1962: 392). A hollowed ladle, filled with gold shavings and plugged with wax, could be exposed to high heat during the course of a fraudulent alchemical "demonstration." When the wax melted and the gold particles secretly dispersed, the illusion would be complete: an otherwise worthless mixture would appear to have been transformed. This same deception was described more than a century earlier in the *Canterbury Tales* of Geoffrey Chaucer. The *Canon's Yeoman's Tale* recounts a canon's tricks on a greedy priest using a "holwe stikke ... stopped with wex" (Chaucer 1913: 256).

Fraudulent gold-making was a cornerstone of alchemical satire for several reasons: the most obvious is alchemy's strained relationship to coinage and currency. If one could produce gold alchemically, was it *as* valuable or *more* valuable than mined gold? Was it even real? Answers were rarely clear. Additionally, anxieties regarding the increasingly global early modern marketplace helped to generate and popularize stock characters peddling "get rich quick" schemes. Alchemists' resemblance to other false professionals in satire – including greedy merchants, two-faced lawyers, and quack doctors – speaks to a baseline of social discomfort (and fascination) around money: spending it, earning it, and making it (Nummedal 2007). Skepticism regarding alchemy's qualifications was also in play. Other artisanal enterprises were guided and upheld by guilds or similarly organized bodies; alchemists had no guild of their own, though individually they might belong to associations of apothecaries, physicians, or metalsmiths through their primary occupations. Guilds tended to stabilize markets, set standards, and support ill members. Without the protections or bona fides guilds provided, alchemists could inhabit a relatively precarious position. However, for some, especially those barred from traditional routes to advancement (poor men, women, members of persecuted religions, and so on), this unfixed status sometimes opened unusual pathways to success. Certain alchemical tracts indicated the breadth of their intended audiences by explicitly addressing not only learned individuals, but also barbers and surgeons, housewives, and all those of poor or "modest means" (Moran 2005: 55). Doctors, inventors, and others sometimes shared the ambiguous territory of the empiric, struggling with similar problem of expertise and authority. In her work on the Royal Society, Barbara Benedict refers to these non-guild "protoprofessional classes" as an "amorphous group resented from both above and below" (2001: 47). This resentment, it seems, stemmed partly from the empirics' inclusion of people who may have emerged in one social level and risen to occupy another.

Skepticism also centered on alchemists' opaque language. Ben Jonson's comic play *The Alchemist*, first performed in London in 1610 by the King's Men, perfectly encapsulates such attitudes. Within the play, three con artists – a servant posing as a sea captain, a trickster playing an alchemist, and a prostitute playing a genteel noblewoman – weave a web of lies via allegorical and metaphorical prose similar to modes of writing employed by real alchemists. Their speeches and performances ironically alter appearance rather than substance. The wittily named Subtle (referencing both his machinations and "subtle bodies," the spirits or natures of substances) plays the alchemist, though he is accused of having previously failed at countless other false professional identities. His assistant and critic, Face, proclaims:

> When all your alchemy, and your algebra,
> Your minerals, vegetals, and animals,
> Your conjuring, cozening, and your dozen of trades,
> Could not relieve your corps with so much linen
> Would make you tinder, but to see a fire;
> I gave you countenance, credit for your coals.
>
> (Jownson 1903: 5)

Subtle's character may have been informed by the real-life exploits of the repeatedly prosecuted London doctor, astrologer, magician, and alchemist Simon Forman, well known during Jonson's lifetime. Forman's illegal practice of medicine gained him the scorn of city physicians, while his love potions and quest for the philosophers' stone gained him infamy and a devoted clientele (Jonson 1903). Jonson himself wrote with more than a cursory knowledge of alchemy. His verses include distinctive terminology and discussions of transmutation's underlying principles. In the second act, as Subtle and Face prepare to con a greedy knight – Sir Mammon – the knight's servant, Surly, accuses the alchemists of false speech. He suggests it is little more than window dressing:

> SUR. What else are all your terms,
> Whereon no one of your writers 'grees with other?
> Of your elixir, your lac virginis,
> Your stone, your med'cine, and your chrysosperm,
> Your sal, your sulphur, and your mercury,
> Your oil of height, your tree of life, your blood,
> Your marchesite, your tutie, your magnesia,
> Your toad, your crow, your dragon, and your panther;
> Your sun, your moon, your firmament, your adrop,
> Your lato, azoch, zernich, chibrit, heautarit,
> And then your red man, and your white woman,
> With all your broths, your menstrues, and materials,
> Of piss and egg-shells, women's terms, man's blood,
> Hair o' the head, burnt clouts, chalk, merds, and clay,
> Powder of bones, scalings of iron, glass,
> And worlds of other strange ingredients,
> Would burst a man to name?

To which Subtle, never at a loss for words, responds:

> SUB. Was not all the knowledge
> Of the Aegyptians writ in mystic symbols?
> Speak not the scriptures oft in parables?
> Are not the choicest fables of the poets,
> That were the fountains and first springs of wisdom,
> Wrapp'd in perplexed allegories?
>
> (Jonson 1903: 152)

Within this verbose theatrical patter there is a grain of truth: the specialized writings of alchemy could indeed cause confusion for nonadepts, or even among the initiated. Reading an alchemical text provided surface access, but understanding its metaphorical language was not a given. Elias Ashmole's mid-seventeenth-century *Theatrum chemicum britannicum*, a compendium of the works of John Dee, George Ripley, and other luminaries of British alchemy, framed alchemical opacity in religious terms: "to the Elected Sons of Art; unto you it is given to know the mysteries of the Kingdome of God; but to others in parables, that seeing they might not see, and hearing they might not understand" (Linden 2003: 223). These alchemical "parables" are precisely what Surly decries: the "tree of life," "red man," and "white woman" of the adepts. At times, allegory acted as a protective form of trade secrecy. Alchemical authors warned their readers that new natural knowledge did not yield itself easily or to the half-hearted: only long study and devotion to the furnace could bring revelation. The thirteenth-century *Radix mundi* (*The Root of the World*), attributed to the friar–philosopher Roger Bacon, decreed that divine judgment had the final say: "If you that are searchers into this science, understand these words ... you are happy, yea, thrice happy; if you understood not what we have said, God himself has hidden the thing from you" (Linden 2003: 122).

Bacon himself appears as an alchemist in Robert Greene's 1589 comic play, *Friar Bacon and Friar Bungay*. Just as the *Radix mundi* warns, a divine power withholds success from Bacon until the final act, due to his moral lapses (including a regrettable interference with the Prince of England's romances). Though the play is fiction, Bacon's works on art, nature, and the emerging sciences were well known in the sixteenth century, and a number of alchemical tracts attributed (spuriously) to Bacon circulated under his name. Bacon was also popularly – though not accurately – associated with occultism and magic. Within Greene's play, gold changes hands and provides temptation, but ultimately it is not plain greed that troubles Bacon. His triumph – the creation of a powerful magical prophesy that assures a new "Golden Age" for England – is nearly derailed by his previous uses of dark magic and forbidden communications with demons. Here even a genuinely skilled adept, in committing a corruption of "true" alchemy, falls short of the integrity required for alchemical perfection (Sadler 1975).

Some alchemical authors might have agreed with the play's themes: their writings not only shared theories, but also worked to "expose" alchemical fraud to a literate public. Heinrich Khunrath, best known for his *Amphitheatrum sapientiae aeternae*, also wrote a vernacular treatise with the lengthy title *Heartfelt Warning and Admonition from a Faithful Devotee of the Truth to All True Enthusiasts of the Natural Transmutational Alchemy, which Necessitates Close Attention Because of the Roguish Tricks of the Wicked Chymists*. It mentions no less than forty-six different types of tricks and cheats, including the hollow stirring rod used in Brant and Chaucer, as well as double-bottomed vessels and bogus glassware. Those so-called alchemists who intended to cheat victims were, in Khunrath's words, "evil ... everyone who sees one should spit at him." The *Chymische Hochzeit Christiani Rosencreutz* (*Chemical Wedding of Christian Rosenkreutz*), published in 1616 and likely penned by the Lutheran utopianist Johann Valentin Andreae, also judged false alchemists by category – first, those fools who deceived only themselves, and second, those who knowingly deceived others. He concludes that the latter "deserve to be sundered from decent folk and severely punished" (Nummedal 2007: 66–70, 162).

Criticizing false practitioners was an intentional strategy employed by alchemists to exude trustworthiness and decency – crafting a serious and respectable alchemical "persona" as counterpoint to popular satires and public doubt. Part of this strategy was the depiction of the "ideal" alchemist. Like the fictional Bacon, a certain nobility of spirit was demanded, but many authors also highlighted the concrete traits of diligence and physical strength needed for laboratory work. The *Magna alchymia* of the charismatic Paracelsian physician Leonhard Thurneisser describes the smelly, cramped, often dirty and hazardous conditions in which alchemists and their assistants might be expected to work. By cataloging (and perhaps hyperbolizing) these dangers, he claims that hardiness, dedication, and meticulousness were requirements for success. As Thurneisser suggests, paired with high morals and devotion to God, as well as vast knowledge of metals, mining and assaying, and alchemical theories, the aspiring alchemist of good character would also be marked by humility, plain words, and simple dress, in contrast to the flashy frauds that the author vilifies elsewhere (Nummedal 2007).

The virtue of curiosity was ascribed to ideal alchemists, for the boundaries of natural knowledge were continually expanding. But curiosity could be a double-edged sword. Those driven by curiosity could create new and useful knowledge, or else be possessed by whimsy and ego (Kenny 2004). Critics of emerging scientific enterprises such as the Royal Society characterized its members as prideful and obsessed with novelty for novelty's sake. Members of the Society were bitingly lampooned in Thomas Shadwell's comic play *The Virtuoso*, which debuted at London's Dorset Garden Theatre in 1676. Nothing was spared: even the famous air-pump experiments of Boyle and Hooke were turned into targets for mockery. The tale centers on the wealthy but hopeless Sir Nicholas Gimcrack, who allows a succession of scientific frauds to empty his bank accounts. In one scene he shares the results of an experimental blood transfusion performed under a questionable "master":

> SIR NIC: But now to return to my transfusion ... I assure you I have transfus'd into a humane Vein 64 ounces Haver du pois weight, from one Sheep. The emittent Sheep dy'd under the Operation, but the recipient Mad-man is still alive.

Yet far from being intrigued, his uncle Snarl laments his nephew's gullibility:

> SNARL: In sadness Nephew, I am asham'd of you, you will never leave Lying and Quacking with your Transfusions and Fools tricks. I believe if the blood of an Ass were transfused into a Virtuoso, you would not know the emittent Ass from the Recipient Philosopher.
>
> (Shadwell 1676: 29–30)

Drawing on themes found in Johnson's *The Alchemist* of sixty years earlier, Shadwell delivers a moral lesson against opportunists who target society's "learned fools" (Benedict 2001: 46). The charges once leveled at alchemists – greed and duplicity, obtuse language, sleights with bogus equipment – look much the same, now laid at the feet of the most famous practitioners of the emerging empirical sciences.

Curiosity's older relationships to insatiable desire and sin continued to influence perceptions of empiricism. Unchecked curiosity could overwhelm morality, as in the *Historia von D. Johann Fausten*, an anonymous work that first appeared at a Frankfurt bookfair in 1587. Among the earliest prose novels written in German, it tells of a brilliant but corrupt student who engages the devil to gain knowledge, wealth, and power. In the end, Faust's damned soul is dragged to hell.[2] The novel narrates battles between lions and dragons, the appearance of a man cloaked in fire with emanating rays, and the devil taking the form of a griffin – all reminiscent of familiar alchemical metaphors. The result of these startling tableaus is Faust's transmutation of gold and silver, presented as an act of witchcraft that violates the authority of nature. As Helen Watanabe O'Kelly concludes, "The narrator of the *Historia* is convinced that these processes can only be the work of the devil" (2002: 31–2).

Faust's story was published within the German territories that cradled Lutheran reform, which might suggest that alchemy was condemned by the faithful. Yet Martin Luther himself praised the language with which alchemists veiled their "truths." In one of the ruminations in his *Table Talk*, a collection first published in 1566, Luther reveals a familiarity with alchemy when he states:

> The science of alchymy I like very well, and indeed, 'tis the philosophy of the ancients. I like it not only for the profits it brings in melting metals, in decocting, preparing, extracting, and distilling herbs, roots; I like it also for the sake of the allegory and secret signification, which is exceedingly fine … for, as in a furnace the fire extracts and separates from a substance the other portions, and carries upward the spirit, the life, the sap … even so God, at the day of judgment, will separate all things through fire.
>
> (Linden 2003: 22)

Luther's reconciliation of alchemical and biblical metaphor stands in counterpoint to Faust and confirms that even religious viewpoints toward alchemy could vary widely.

Though damning responses to fraudulent alchemists appear throughout literary fiction and theater, as well as in the writings of alchemists themselves, this censure does not seem to have found a similarly broad expression in prints and painting. In the visual arts, resentments and accusations of fraud against alchemy were largely transformed into vignettes of sympathy, head-shaking laughter, and pity. While Thomas DaCosta Kaufmann (1997) has suggested a lasting thread of antialchemical thought among early modern artists and their communities, particularly in the Dutch Republic and other Northern regions, most images of alchemy remain markedly less antagonistic toward alchemists than their counterparts in text.

FOOLS, SCHOLARS, AND MASTERS IN PRINT

Pieter Bruegel the Elder's "The Alchemist," engraved after 1558 by Philips Galle and published in Antwerp by Hieronymus Cock (see Figure 3.2), is the most often-cited origin point for the popular imagery of the alchemist that flourished in Northern Europe during the following century (Orenstein 2001). The print shows an alchemist in ragged clothes whose crucible consumes his family's last coins. His wife holds an empty purse while his three children play in an empty cupboard, suggesting hunger and privation. At center, a fool in a donkey-eared cap works a hand bellows. These futile actions are observed by a scholarly figure seated at a lectern, whose text warns *al-ghemist* – literally "all is shit (manure)," a vulgar Dutch pun (Principe 2013: 183). A copy after this print, engraved in the first quarter of the seventeenth century by Claes Jansz. Visscher II, bears a more bluntly damning inscription:

> The alchemist, very much reviled, seeks fine gold and treasures. What he had is now gone. He takes the four elements, yet gains nothing. His labors spoiled and fortunes ruined, this poor man's life ends at the almshouse.

This later print also adds numerous inscriptions in Latin and Dutch. The oversized distillation vessel is labeled "Aqua Magistralis," or "teacher's water," a remedy believed to treat syphilis.[3] The scholarly man at right is "Doctor Loshooft," literally "loose-head." His stool is labeled "Vanitas," a warning against the search for worldly gold and glory. Even the materials piled beneath the furnace are "lutum sapientia," or "mud wisdom," a crude joke on the alchemical search for purified matter. The impact of these prints was widespread, and elements of Bruegel's design reemerged numerous times in Netherlandish painting of the seventeenth century, where alchemists became a remarkably popular motif.

Though influential, Bruegel's print has been overemphasized in the history of art as a blueprint or origin point for *all* images of alchemy. This interpretation roots the very representation of alchemy in satire. However, this print was neither the first nor the only representation of alchemy available to collectors or to other artists. Also circulating were other sixteenth-century images produced by the Flemish artist Jan van der Straet, better known to his international patrons as Johannes Stradanus, as well as illustrations from alchemical texts and early encyclopedias. These offered alternative visions of a legitimate and beneficial alchemy that produced useful goods and performed natural philosophical inquiry. Bruegel's and Stradanus' prints in interaction help to demonstrate that alchemy was not understood as a singular moral or social absolute, but as a set of practices and processes that could be used or misused.

Half a century before Bruegel's image, depictions of alchemists as scholars appeared in northern encyclopedic texts. Among the first was a small woodcut of an alchemist tending a fire in the *Margarita philosophica* (*Philosophical Pearl*), an encyclopedia published in Freiburg in 1503. The woodcut's simple design shows an alchemist in fine robes holding bellows at a fire, flanked by a distillation vessel topped with an alembic. The alchemist's fashionable clothing – a robe with slit sleeves and a scholar's cap – is nearly identical to another illustration of a studious mathematician from the same text. The alchemist of the *Margarita philosophica* was a figure of both scholarly authority and hands-on work; later images would emphasize alchemists' manual skills in distillation and metallurgy. Lazarus Ercker's text on mining, the *Aula subterranea* (*Subterranean Hall*), first published in Prague in 1574, presented alchemy in the context of a busy workshop. Several woodcuts representing an assaying laboratory show workers tending to furnaces and vessels. Traditional divisions of labor are present: menial tasks such as grinding with a mortar and pestle are shown performed by youthful assistants, while more finely dressed adults are shown as supervisors, gauging the results of the process. Rather than hermetic scholars, these figures are masters at the center of a dynamic commercial enterprise.

Illustrations of specific equipment and apparatuses were also features of alchemical texts themselves. One 1545 Latin translation of the works of Pseudo-Geber contains detailed images of furnaces, glassware, and distillation apparatuses, including a woodcut of two long-necked alembics and cucurbits and an accompanying water bath (Figure 8.1). Details of rising flames and dripping water emphasize the practical applications of such illustrations as sources of guidance for setting up and maintaining laboratory functions. Systematically categorized diagrams of the many possible types and shapes of alembics and other glassware also appear in the 1597 *Alchemia* of Andreas Libavius, a German university lecturer and alchemist. The work was an attempt to compile collected texts of preceding generations, and it reveals Libavius' desire not only to educate aspiring adepts, but to formalize and legitimize alchemy within the

FIGURE 8.1 Pseudo-Geber, illustration of water bath apparatus from the *Summa perfectionis magisterii*. Venice, 1542. Othmer Library, Science History Institute, Philadelphia.

established framework of the university (Moran 2007). Paintings of alchemical workrooms, which I will discuss below, often include representations of printed book illustrations that have been torn out and pinned up on laboratory walls beside hand-drawn diagrams and notes. Artists' inclusions of such details suggest that the purpose of these technical illustrations – to help guide and educate in the setup of experiments – was widely understood.

Detailed depiction of apparatus could act as a visual argument for alchemy's technological achievement and economic power. This is evident in Johannes Stradanus' *Distillatio* (see Figure 5.3), part of his *Nova reperta* (*New Discoveries*), a print series celebrating landmark inventions and novelties of the era, ranging from the "discovery" of the New World to the development of the printing press itself. Designed by Stradanus while in the employ of the Medici court in Florence, engraved by Jan Collaert and published around 1590, the *Nova reperta* was an essentially modern project that blended older trade imagery and pictures of "curiosities" with of-the-moment technological innovation (Markey 2012: 392–3). *Distillatio* shows a seated alchemist examining a heavy text as his assistant points to an open page. The workshop is spacious and grand and allows the scene to show multiple stages of the distillation process. In front, a young assistant grinds herbs (likely for medicinal use) and grain (for alcohol) in a large mortar and pestle attached to a spring-arm. At the center an exaggeratedly massive still empties vapor into long-necked retorts, and at right a "hooded" still or *Rosenhut* sits above a smoking furnace. The obvious prosperity and productivity of the workshop demonstrates what John Read has called "ordered and affluent activity" (1947: 66–8). The engraving's caption praises the workshop's products: "In the juice of the fire, by art, these bodies / become a stream of water, pure and most powerful."

Stradanus' patron Francesco I, Grand Duke of Tuscany, also commissioned paintings of laboratories and workshops for his private studio. In 1570, Stradanus produced *Il laboratorio dell' alchimista* (*An Alchemist's Laboratory*), a picture that anticipates his *Distillatio*. The scene shows his patron in the role of an alchemist's assistant: the Grand Duke appears as a humble adept heating liquid in a pan under the watchful eye of a scholar. The latter figure closely resembles the seated alchemist of *Distillatio* and was likely modeled on Francesco's real-life teacher Josef Goodenhuyse, a Flemish natural philosopher. The painting was installed alongside other works presenting recent material innovations and industries ranging from wool mills to glassworks. In both painting and print, Stradanus' compositions emphasize the duality of theory and practice through a subtle division: one side of each image deals with the "known" – in the form of printed books and the workshop master – while the other half focuses on the "new" – in the form of more experimental labor (Cerruti et al. 2008: 725).

Perhaps the most succinct visual document of alchemy's paradoxical status in print comes in the form of an engraved illustration from Christoph Weigel's 1698 book of trades, the *Abbildung der gemein-nützlichen Haupt-Stände ...* (*An Illustration of Common and Useful Trades*). Weigel, together with father-and-son engravers Jan and Caspar Luyken, produced over two hundred images of "*alle Künstler und Handwerker*" (all artisans and tradesmen), including an image of two alchemists (Figure 8.2) titled "Der Alchymist oder Goldemacher" (Lis and Soly 2012). At first glance, "alchemist or gold-maker" reads as two interchangeable names for the same profession. Yet this interpretation is abruptly reversed by the brief epigram below it, which praises alchemy for its improvements of human life – with the exception of its suspect branch, *chrysopoeia*. The text praises alchemy's utility while cautioning against vainglory and greed: "Such artifice is necessary / to prepare medicines / and to help nature / But when you seek

FIGURE 8.2 Jan and Caspar Luyken after Christoph Weigel, "Der Alchymist oder Goldemacher," ca. 1698. Engraving. Wellcome Collection, London CC BY.

gold alone / so will vanish with the smoke / honor, laughter, money, and mercury." Evidently, early modern audiences were open to evaluating alchemy's diverse practices on their individual merits. This engraving, produced in the final years of the seventeenth century, crystallizes the ever-shifting balance of anxiety, suspicion, curiosity, and admiration that wove through the era's depictions of alchemy, and it challenges our long-held contemporary reliance on Bruegel as sole or *prima* artistic *materia*.

LABORATORIES IN SEVENTEENTH-CENTURY PAINTINGS

Even when painters of the seventeenth century did choose to borrow from Bruegel's sixteenth-century composition, they did so selectively. Rather than copying the false scholar, or the fool in a jester's cap, most artists adopted only the alchemist and his family. Overall, the anxieties about alchemical fraud that appeared so often in literature and theater appeared very little in painting. Instead, most artworks foregrounded alchemical *folly* – usually in the form of a ragged alchemist and an impoverished family. This choice provided greater pathos and emotional immediacy, in keeping with broader trends in art that shifted away from overt didacticism and toward greater naturalism. Artists who produced alchemical works echoing Bruegel include the Dutch painters Adriaen van de Venne, Adriaen van Ostade, and Richard Brakenburgh and the Flemish artist Mattheus van Helmont, among others. Yet some of their peers – including Thomas Wijck and Cornelis Bega in the Dutch Republic and David Teniers in the southern Netherlands – challenged the Bruegelian pictorial tradition and created images that explored alchemy's utility and benefits, as well as its connections to art and artisanal work.

These diverse scenes model varied cultural attitudes toward alchemy, from respectful curiosity to outright mockery. Importantly, these varied attitudes often find outlet in different works by the same painter. An artist's willingness to paint alchemical satire did not mean they were unwilling to paint alchemy seriously. Certain artists, such as Heerschop and van Helmont, produced scenes of both foolish and respectable alchemists. Interpreting these pictures therefore relies on their subtle treatments of signs and conventions, rather than merely on their subject matter. Painted as a peasant, the alchemist dreaming of gold appears foolish. Painted as a craftsman in a busy workshop, he becomes the subject of admiration and interest. Multivalent and sophisticated, many of these works play mockery against sympathy, fascination against doubt, inviting the viewer's own reciprocal curiosity and analysis. Paintings are powerful tools for carrying and shaping social values, yet they can also be understood as exercises in wit, play, and experiment (Westermann 1997).

FIGURE 8.3 Adriaen van de Venne. *Rijcke-Armoede*, ca. 1630–1632. Oil on panel. Science History Institute, Philadelphia.

Wit and experiment are central to Adriaen van de Venne's 1630 image of an alchemist at the fire (Figure 8.3), titled *Rijcke-Armoede* (*Rich Poverty*). Van de Venne's use of a sepia-toned *brunaille* for such a humble subject is playful: *grisaille* or *brunaille* palettes evolved from the decoration of altarpieces and were originally employed for elevated and serious subject matter. Elite connoisseurs appreciated the special skill of working in a limited palette, and van de Venne spent much of his life working for such a clientele in the Dutch capital of The Hague. The coloring of *Rijcke-Armoede* – a sophisticated method for a "low" satirical subject – added a clever, humorous quality (Franits 1993: 88). "Rich poverty" may imply that the alchemist, shown with crucible and tongs, is a fool whose work with gold ironically amasses no wealth. Yet other elements suggest alternative interpretations. Though his circumstances are shabby, the alchemist's equipment is specialized and portrayed with a degree of accuracy. At right, the bottom vessel or *cucurbit* topped with a long-stemmed alembic is of a type that frequently appeared in illustrations for alchemical treatises on distillation and vaporization (Principe and DeWitt 2002: 17). Van de Venne's education and facility with Latin and his involvement with the family's successful publishing business (Franits 1993: 87–8) suggest he may have encountered alchemical ideas within intellectual and courtly circles. Thus "rich poverty" might also denote knowledge that is undervalued or has not yet yielded a return. This common professional complaint – undervalued skill – would not be unique to alchemy and could provide a point of humanizing sympathy. Other objects within *Rijcke-Armoede* are drawn directly from Bruegel, including the oversized distillation apparatus to the left, the makeshift table resting on a barrel, and the pitiable open-handed gesture of the alchemist's wife. Yet van de Venne transforms Bruegel's comic lesson into a more intimate scene of a family's anguish.

Alchemy's impact on family life offered painters an outlet for humor as well as pathos. The painter Richard Brakenburgh was a resident of the Dutch city of Haarlem, which produced alchemical genre paintings in particularly great volume. Brakenburgh's *An Alchemist's Workshop with Children Playing* (Figure 8.4) depicts an alchemist as the irresponsible leader of a teeming household whose wife pleads for consideration and whose children play at cooking and eating beside empty cupboards. A young boy with a copper pan and a pair of

FIGURE 8.4 Richard Brakenburgh, *An Alchemist's Workshop with Children Playing*, ca. 1670–1680. Oil on canvas. Science History Institute, Philadelphia.

bellows on the floor imitates the work of his father, a comic detail that also indicates the alchemist's failure as a parental model. As Lawrence Principe has noted, Brakenburgh's composition simultaneously quotes and transforms Bruegel by replacing the fool at center with the alchemist's own son (Principe 2012). The work is simultaneously an amusing, common Dutch scene of a "disorderly household" and a sentimental, heavily moralizing picture of disappointed family hopes. The latter tone is quietly emphasized by devotional objects placed around the room, including a panel painting of the Madonna and Child on the rear wall and a crucifix sitting on a shelf at right. The contrast of Christian ideals of motherhood and family sanctity makes clear the alchemist's moral failings as a provider (Principe and DeWitt 2002).

Hendrick Heerschop's *The Alchemist's Experiment Takes Fire* (see Figure 5.4) shows a metallurgical experiment gone wrong, but the alchemist's shocked expression and flying fragments of glass provoke laughter more than moral outrage. The explosion is paired with a scatological pun glimpsed through a doorway, where a mother is shown wiping her child's soiled bottom. The widespread popularity of this pun in Dutch art is underscored by Johan de Brune's 1624 *Emblemata*, where a similar scene of mother and child is accompanied by the dour message, "what is life, but shit and stink?" (Schama 1987: 481–2). Literal bodily waste mocks the alchemist's wasted actions. Yet Heerschop's crude and playful joke at alchemy's expense was only one among his various approaches to alchemical subject matter: in *An Alchemist and His Assistant* (Figure 8.5) he eschews humor to present a scholarly alchemist interrupted in study by a workman. The latter's rough clothing suggests he is engaged in dirty and demanding laboratory work, while the contents of the alchemist's small studio indicate mental labors. The globe, mortar and pestle, and apothecary's jars are conventional elements of a painted alchemist's mise-en-scène, but a large open book resting on the floor also displays two illustrations – one of a plant specimen, one of a human figure – that suggest special facility with botany and anatomy, likely relating to medicine. Here, respectability and expertise define the alchemist, rather than folly or neglect.

To the south, the Flemish painter Mattheus van Helmont (no known relation to the Flemish chymist J.H. van Helmont of a generation earlier) produced a similarly mixed collection of alchemical scenes. His *An Alchemist at Work* (Figure 8.6) shows a laboratory in an exaggerated state of disarray and chaos, strewn with waste materials: likely an echo of the "mud wisdom" pictured by Bruegel. The workshop's barn-like atmosphere includes piles of straw, twigs, branches, and broken charcoal. As the alchemist contemplates a glass vial, adjusting his spectacles, the many books nearby appear to offer no help. As Jan Bialostocki (1988) has noted, early modern depictions of books do not always signal wisdom or scholarship. Rather, abundant texts paired with foolish readers instead signal a lack of learning, an overreliance on dogmatic text over critical

FIGURE 8.5 Hendrick Heerschop, *An Alchemist with His Assistant*, ca. 1660–1680. Oil on canvas. Science History Institute, Philadelphia.

thought, or a gap between *reading* and *knowing* that no amount of written words could bridge. This parallels the warnings of alchemical writers such as Ashmole, who stated that while many aspire to alchemy, most would remain among those who "seeing ... might not see, and hearing ... might not understand." A second scene by van Helmont, *The Alchemist* (Figure 8.7), retains a few threads of straw but shifts modes entirely to present the alchemist as a learned and worldly scholar whose work borders other sciences as well as the arts. At left, a large

FIGURE 8.6 Mattheus van Helmont, *An Alchemist at Work*, ca. 1650–1680. Oil on canvas. Science History Institute, Philadelphia.

FIGURE 8.7 Mattheus van Helmont, *The Alchemist*, ca. 1650–1680. Oil on canvas. Science History Institute, Philadelphia.

écorché figure reminds viewers that both artists and *iatrochemists* (alchemists engaged in chemical medicine) studied the inner workings of the body. A violin hung on the rear wall and a skull overturned on the table suggest musings on harmony and mortality. The presence of a large furnace and workers gathered near an anvil to the rear of the picture does not dispel the sense of poetic or creative contemplation suggested by the rest of the scene.

Works by the most prolific Flemish painter of alchemical laboratories, David Teniers the Younger, likewise connect alchemy not only with chemical work and productivity, but also with deeper explorations of inventive genius. Teniers' thematic treatments of alchemy align more closely with the attitudes found in works by Stradanus rather than the comic pauper-alchemists of Bruegel and his followers. Teniers headed a sizeable studio and produced paintings both for the thriving Antwerp market and a number of royal patrons; by 1647, Teniers was in the service of the Archduke Leopold Wilhelm in Brussels, and most of his alchemical works date from this period through the 1680s (Davidson 1979). Leopold's first cousin Rudolf II was well known for his empirical interests, as well as for his patronage of experimental painters such as Giuseppe Arcimboldo and Bartholomeus Spranger (Marshall 2009). Like Stradanus, Teniers was likely introduced to alchemy in a cosmopolitan court environment.

Teniers' *Alchemist with Book and Crucible* (Figure 8.8) shows a modest workroom equipped with a large hearth. In the foreground the alchemist stands

FIGURE 8.8 David Teniers the Younger, *Alchemist with Book and Crucible*, ca. 1640–1670. Oil on panel. Science History Institute, Philadelphia.

thoughtfully reading from a book while tending a crucible over a pile of hot coals. His young assistant looks on with eager attentiveness, while other workers are gathered around a bench to the rear. While this scene lacks the scale and bustle of Stradanus' *Distillatio*, the room's orderly operation suggests the alchemist's authority and mastery. Teniers' alchemist wears the robes of a scholar, paired with heavy boots or gaiters to shield against the furnace's heat. His slightly distracted air and the simplicity of his dress may be signs of his complex mental and imaginative capacities, as in representations of contemplative scholars and artists of the same era.[4] Intriguingly, a glass gazing-ball appears in the top center of the painting, suspended from the room's ceiling. In Dutch and Flemish art, gazing-balls and reflective "bubble" glasses often symbolized the brevity of life (referring to the aphorism "*homo bulla*," or "man is a bubble"); yet they were also utilized as self-conscious references to representation itself, a means to explore imitation and reflection – and the very nature of art and illusion (Martin 2011). Teniers' inclusion of the gazing-ball may speak to the alchemist's age, but also implies a connection between the mimicry and emulation performed by both artist and alchemist. Considering this detail, Teniers' choice to depict *himself* as an alchemist in at least one self-portrait is especially resonant. Such a self-portrait would recall for his viewers the many popular alchemical pictures produced by his workshop, yet it also speaks to his sophisticated understanding of the sympathies between two disciplines that explored the natural world in pursuit of perfection. Both alchemy and art were referred to as "apes of nature," their practices imitating nature much as apes and monkeys were believed to imitate human activities. This connection was made literal in Teniers' many *singeries*, images that replaced human actors with apes and monkeys for comedic results. A lost painting by Teniers, reproduced by the eighteenth-century French engraver Pierre François Basan, *Le Plaisir des Fous* (*The Pleasure of Fools*), shows a gaudily dressed monkey-alchemist "puffing" with bellows before the fire (Figure 8.9). It may seem like cutting mockery, yet Teniers also produced numerous *singeries* of monkey-artists wielding paintbrushes, showing his playful approach even to the discipline that brought him fame and fortune (Davidson 1979).

The art–alchemy connection also underlies works by the Dutch painter Cornelis Bega. At first glance, *The Alchemist* of 1663 (Figure 8.10) appears to be another scene of failure and squandered resources – an alchemist alone in the ruin of his workshop. Yet its emphasis is on work rather than on pathos and poverty. The alchemist's equipment bears signs of a long and tumultuous process: broken glass and ceramic vessels lie in a heap. Yet his labors may not have been in vain. The alchemist's fallen stocking and bared knee can conjure up images of dissolute living, yet they also evoke stories of the "wandering adept,"

FIGURE 8.9 *Le Plaisir des Fous*, Pierre Francois Basan after David Teniers the Younger. Science History Institute, Philadelphia.

FIGURE 8.10 Cornelis Bega, *The Alchemist*, ca. 1663. Oil on panel. Getty Museum, Los Angeles/digital image courtesy of the Getty's Open Content Program.

skilled alchemists whose unremarkable appearance hid seemingly supernatural talents – just as base lead was believed to hold the potential of transmutation into gold (Principe and DeWitt 2002). The presence of vivid red matter also signals the possibility that this alchemist may have achieved the triumphant final stage of the "great work." The philosophers' stone was described as a red substance or elixir produced in the final "red stage" or *rubedo*. The white chunks of matter loosely wrapped in paper and cloth at bottom right may also represent an earlier "white" or *albedo* stage in the same process (Principe 2012).

FIGURE 8.11 Cornelis Bega, *The Alchemist*, ca. 1663. Oil on canvas, mounted on panel. From the Collection of Ethel and Martin Wunsch/National Gallery, Washington, DC.

A second scene by Bega, also entitled *The Alchemist* (Figure 8.11), again places emphasis on the products of the laboratory. Though once again the alchemist is surrounded by broken vessels, he is not dirty or ragged, but wears a workman's protective vest and gaiters. At upper right, a large earthenware apothecary's jar bears the label "SILICON," likely a fragment of *basilicon*, a common unguent applied topically for pain. Bega's choice is particularly interesting in light of basilicon's ingredients, nearly all of which were substances well known to painters. An influential medieval medical tract, the *Inventarium sive chirurgia magna* (*The Inventory* or *Great Work of Surgery*) written by Guy de Chauliac around 1363 and republished through the seventeenth century, provides twelve different recipes for basilicon. One lists *cire* (wax), *mastic* (a resin used in varnishing), *verd de gris* (verdigris, a green pigment composed of copper salts, sometimes used as a drying agent), *terebinthine* (turpentine), *litharge* (a white pigment, lead oxide), and *galban* (another resin) as ingredients (De Chauliac 1580: 674–6). These materials were ubiquitous in painter's workshops. The prominence of the jar and its clear legibility indicate Bega's desire to highlight the substance within, perhaps because of its obvious utility, or more likely because he was conscious of its relationship to his own materials. As the grandson of a painter, son of a silver- and goldsmith, and nephew of an engraver and silversmith (Van Thiel-Stroman 2006), Bega was perfectly positioned to appreciate the importance of shared chemical, metallurgical, and artistic traditions.

Artisanal processes – messy, laborious, and creative – are also centered in the laboratories of Thomas Wijck, a fellow Haarlem painter who produced several dozen scenes of alchemists at work. His *Interior with an Alchemist* (Figure 8.12) depicts alchemical clutter in lavish detail. Apothecary jars, copper basins, books, jars, glassware, and tools are piled to the left and right; to the rear of the workshop a young assistant works with a large mortar and pestle. Every visible surface – chairs and cabinets, the alchemist's desk, windowsill, and floor – is covered in small vials, bundles, and paper folios. The clutter is insistently foregrounded, suggesting that it is key to understanding the picture as a whole. As noted in the introduction to this chapter, most interpretations of alchemical mess have been unequivocally negative; Jane Russell Corbett writes, "It is difficult to respond to the general disorder of these workrooms as anything other than a negative comment" (2005: 254). Yet the laboratories in Wijck's paintings were considered "ingenious" by his biographers: the prominent seventeenth-century art critic Arnold Houbraken called their details "wittily painted" and "artfully suitable." Houbraken also listed specific objects typically shown by Wijck: "furnaces, crucibles, pans, glasses … a multitude of tools" (1976: 16); a detail that suggests other period viewers would have been equally able to identify vessels and apparatus among general clutter.

The artfulness and "suitability" of Wijck's pictures lay partially in his naturalistic representations of real vessels. And yet the manner of those representations also speaks to the nature of alchemical work as an investigative process. Though it produced many useful goods for the market, profit was not alchemy's only aim. Many of its practitioners were engaged in seeking natural secrets – things that might upend knowledge of matter, and in turn knowledge of the universe. This work was not unlike the creative labors of artists, whose

FIGURE 8.12 Thomas Wijck, *Interior with an Alchemist*, ca. 1660–1677. Oil on panel. Wellcome Collection, London CC BY.

imaginations or *fantasia* were at times considered dangerous for their unbridled potential (Cole 2002). A new interpretation of the mess on view within Wijck's *Interior with an Alchemist*, therefore, might take into consideration the alchemist's – as well as the artist's – ongoing creative endeavors. A book left open implies a return to reading or an unfinished thought. A pile of cracked vessels indicates repeated tries at mastery, each attempt yielding new data. The scene communicates the long duration of process and the unfixed quality of alchemy's interrogations of nature.

FIGURE 8.13 Thomas Wijck, *Alchemist and Family*, ca. 1660–1670. Oil on canvas. Science History Institute, Philadelphia.

At times, Wijck's images also modeled alchemy as a setting for domestic harmony. The *Alchemist and Family* of about 1660 (Figure 8.13) shows the alchemist's wife at center, preparing food while her children watch and imitate. The alchemist, seen through a stone alcove at left, sits in study. The diligence of the alchemist's wife and the positively imitative behavior of their children challenge the pictorial traditions originated by Bruegel and continued by painters such as Brakenburgh. Wijck transforms neglected wives and hungry children into a pleasant scene of family harmony, with the alchemist as responsible paterfamilias and alchemy as a respectable profession. In reality, many alchemical practitioners made use of whatever facilities were available to them – including kitchens, barns, studies, and sheds. Practically speaking, domestic and alchemical labor were frequently intertwined. Alchemical texts were also replete with allusions to mundane domestic tasks: the *magnum opus* or "great work" was sometimes referred to as the *opus mulierum* or "women's work." This mode paired phases of a specialized, secret process with common household tasks: distillation with washing and boiling, fermentation with cooking and food preparation, and so on. Margaret Cavendish, a seventeenth-century English duchess and natural philosopher, declared in her writings that nature itself is a "good housewife," and that the diligence of women (like true adepts) marked their suitability for alchemical endeavors (Archer 2010: 195). Women's participation in alchemical enterprise was not merely allegorical, though it frequently went undocumented. Remedies and tonics, as well as dyes and other chemical products, were often manufactured by women, and most artisanal trades relied upon the skills and support of female family members. While Wijck's picture does not directly portray women's involvement with natural philosophy, his emphasis on domestic labor reminds us of the necessity of that work, a detail often obscured in histories of experimentation.

CONCLUSIONS

The majority of the images discussed in this chapter represent alchemical practice (i.e. relatively naturalistic depictions of the laboratory and its inhabitants at work). Yet there is another side to the representation of alchemy: the representation of alchemy's *ideas*, its abstractions and theoretical inquiries. Alchemists were urgently concerned with such projects and often included symbolic diagrams in their treatises in addition to more straightforward images of tools and apparatus. These diagrams ranged from cosmologies that organized the universe according to alchemical elements and affinities, to deliberately coded emblems of stages in the "great work," to mnemonic devices and visualizations of the alchemist's path toward enlightenment. The lavish double-page illustrations of Heinrich Khunrath's *Amphitheatrum sapientiae aeternae* beautifully demonstrate both the collaboration of alchemists and artists (in this case, Khunrath and the engraver Jan Diricks van Campen) and the lengths of symbolic layering to which alchemical diagrams might go. Plate 3 (Figure 8.14) shows a hypnotic glimpse through a massive cave tunnel ringed with inscribed stone tablets. On the steps below, miniscule alchemists traverse the

FIGURE 8.14 Heinrich Khunrath, Plate 3 from *Amphitheatrum sapientiae aeternae* (*Amphitheater of Eternal Wisdom*). Hanau, Germany: Wilhelm Antonius, 1609. Othmer Library, Science History Institute, Philadelphia.

dark passageway, seeking the illuminated "door to eternal wisdom" glimpsed at the far end.

While images such as Khunrath's were evocative and inspired deep contemplation and study, they were rarely adapted by artists for independent works. Their relative impenetrability to the casual viewer – by design – and their symbolic ecstasies were at odds with seventeenth-century trends toward greater naturalism in compositions and the adoption of "modern" subjects, such as scenes of city life and new trades. Those artists (and authors) who did depict alchemy reveal by their selections and omissions quite telling information about their own preoccupations and perceptions. It is amusingly fitting that early modern authors and playwrights should have focused attention primarily on alchemists' secretive or allegorical language, while printmakers and painters attended to questions of materiality and visualized parallels between alchemical and artisanal disciplines. The subjective nature of these depictions reminds us that plays, books, and pictures are constructed spaces that reveal as much (or more) about their social and artistic contexts as they do about actual alchemical practice. While comic plays about alchemists are rarely mistaken for documentation of alchemical work, printed and painted laboratories are occasionally treated as such. Instead of seeking alchemical "truths" in pictures, we can savor the ways that they model attitudes and conventions; the ways in which they grant access to modes of thinking that have since altered or vanished. Authors and artists were not passive copyists of the alchemical world, but active translators of it. Their works shaped public perceptions of alchemy and aided in either discrediting or promoting its practitioners, and they perform those same tasks still.

There were many alchemies – diverse methods and practices that spanned medicine, philosophy, metallurgy, cosmology, and more – and there were equally many ways of representing their practitioners in the arts. Alchemy yielded answers as well as new questions. It broke glassware and consumed coal and spurred arguments and occasionally returned metaphorical gold. The legendary messiness of alchemy's experiments, which artists captured with so much delight, was both literal and figurative, for alchemy readily spilled over and across the boundaries of whatever discipline it touched. That these texts and pictures should still speak to us today with so many competing voices is a pleasant testament to that fact.

NOTES

CHAPTER 4

1. Other late seventeenth-century figures much like Kunckel include Johann Joachim Becher and Johann Rudolph Glauber.
2. Von Hohberg wrote that he was not prepared, however, to deny that transmutation was possible.
3. The title of Clajus' book, *Altkumistica*, was a pun, which equated alchemy with "old cow shit."
4. No full census to my knowledge as yet exists. As an indication, the database VD17 lists 1,554 imprints for the seventeenth century that catalogers have assigned to the subject areas of either alchemy or chemistry or both.

CHAPTER 5

1. The manuscript, now in the Spencer Research Library at the University of Kansas, is labeled Pryce MS E1.
2. It was not until the eighteenth century that gunpowder chemists standardized the most effective mixture: around six parts saltpeter to one part sulfur and one part charcoal.

CHAPTER 8

1. Like the majority of laws passed against the practice of alchemy, Henry IV's prohibition against transmutation had largely centered on the possibility of a devalued currency and the potential for financial fraud.
2. While the novel was a work of fiction, a historical figure of similar name was thought to have lived in Saxony, and was mentioned negatively several times by early sixteenth-century Lutheran reformers, including Philip Melanchthon.

3 "*Aqua magistralis pro iis qui occulté morbo Gallico curari desiderant*," listed in the index of a 1667 publication compiling the works of Zacuto Lusitano. *Zacuti Lusitani, medici, & philosophi praestantissimi* ... Lugduni: Sumptibus Ioannis Antonii Huguetan, filij, & Marc Antonii Rauaud., M. DC. XLIX.
4 Rembrandt's "coarse, careless" appearance in self-portraits was discussed by contemporaries as praiseworthy evidence of his total immersion in the studio (Chapman 1990: 96–7).

BIBLIOGRAPHY

Abbri, Ferdinando. 2000. "Alchemy and Chemistry: Chemical Discourses in the Seventeenth Century." *Early Science and Medicine*, 5, 214–26.
Agricola, Georgius. [1556] 1557. *De re metallica*. Basel: Emanuel König.
Agricola, Georgius. 1557. *Vom Bergkwerck XII Bücher darinn alle Empter, Instument, Gezeuge unnd alles zu disem handel gehörig, mitt schönen figuren, vorbilder und klärlich beschriben seindt*. Basel: Froben.
Agricola, Georgius. [1912] 1950. *De re metallica*, Herbert Clark Hooker and Lou Henry Hoover (trans.). New York: Dover Publications.
Agrippa, Heinrich Cornelius. [1531–3] 1898. *Three Books of Occult Philosophy*. Willis F. Whitehead (ed.). Chicago, IL: Hahn and Whitehead.
Anderson, Barbara C. 2014. "Evidence of Cochineal's Use in Painting." *Journal of Interdisciplinary History*, 45: 337–66.
Anon. 1550. *De alchimia opuscula complura veterum philosophorum*. Frankfurt am Main: Cyriacus Jacobus.
Archer, Jayne E. 2010. "Women and Chemistry in Early Modern England: The Manuscript Receipt Book (c.1616) of Sarah Wigges." In Kathleen P. Long (ed.), *Gender and Scientific Discourse in Early Modern Culture*. London: Routledge.
Arntz, Helmut. 1975. *Weinbrenner: die Geschichte vom Geist des Weines*. Stuttgart: Seewald.
Arrizabalaga, Jon, John Henderson, and Roger French (eds). 1997. *The Great Pox: The French Disease in Renaissance Europe*. New Haven, CT: Yale University Press.
Ash, Eric. 2004. *Power, Knowledge, and Expertise in Elizabethan England*. Baltimore, MD: Johns Hopkins University Press.
Aubrey, John. 1898. *"Brief Lives," Chiefly of Contemporaries*. Andrew Clark (ed.). Oxford: Clarendon.
Bacon, Francis. 1605. *The Twoo Bookes of Francis Bacon. Of the Proficience and Advancement of Learning, Divine and Humane*. London: for Henrie Tomes.
Bacon, Francis. 1627. *Sylva Sylvarum, or, A Naturall Historie*. William Rawley (ed.). London: William Lee.
Bacon, Francis. 1659. *New Atlantis …* London: Thomas Newcomb.

Bacon, Francis. 1960. *The New Organon and Related Writings*. Fulton H. Anderson (ed.). Indianapolis, IN: Bobbs-Merrill.

Bakewell, Peter. 1983. *Miners of the Red Mountain: Indian Labor in Potosí, 1545–1650*. Albuquerque: University of New Mexico Press.

Bakewell, Peter. 1984. "Mining in Colonial Spanish America." In Leslie Bethell (ed.), *The Cambridge History of Latin America, Vol. II, Colonial Latin America*. Cambridge: Cambridge University Press.

Baldwin, Martha. 1993. "Alchemy and the Society of Jesus in the Seventeenth Century: Strange Bedfellows?" *Ambix*, 40: 41–64.

Barrera-Osorio, Antonio. 2006. *Experiencing Nature: The Spanish American Empire and the Early Scientific Revolution*. Austin: University of Texas Press.

Bäumel, Jutta. 2004. "Electoral Tools and Gardening Implements." In Dirk Syndram and Antje Schemer (eds), *Princely Splendor: The Dresden Court 1580–1620*. Milan: Electa.

Becher, Johann Joachim. 1669. *Moral Discurs von den eigentlichen Ursachen dess Glücks und Unglücks ...* Frankfurt am Mayn: In Verlegung Johann David Zunners.

Béguin, Jean. 1669. *Tyrocinium Chymicum or, Chymical Essays, acquired from the Fountain of Nature and Manual Experience*. London: for Thomas Passenger.

Belfanti, Carlo Marco. 2004. "Guilds, Patents, and the Circulation of Technical Knowledge: Northern Italy during the Early Modern Age." *Technology and Culture*, 45: 569–89.

Benedict, Barbara M. 2001. *Curiosity: A Cultural History of Early Modern Inquiry*. Chicago, IL: University of Chicago Press.

Bentancor, Orlando. 2007. "Matter, Form, and the Generation of Metals in Alvaro Alonso Barba's *Arte de los metales*." *Journal of Spanish Cultural Studies*, 8: 117–33.

Benzenhöfer, Udo. 1997. *Paracelsus*. Reinbek bei Hamburg: Rowohlt.

Beretta, Marco. 2014. "Material and Temporal Powers at the Casino de San Marco (1574–1621)." In Sven Dupré (ed.), *Laboratories of Art: Alchemy and Art Technology from Antiquity to the 18th Century* (Archimedes, vol. 37). Cham: Springer.

Berger, Peter L., and Thomas Luckmann. 1966. *The Social Construction of Reality: A Treatise in the Sociology of Knowledge*. Harmondsworth and New York: Penguin.

Bernardoni, Andrea. 2014. "Artisanal Processes and Epistemological Debate in the Works of Leonardo Da Vinci and Vannoccio Biringuccio." In Sven Dupré (ed.), *Laboratories of Art: Alchemy and Art Technology from Antiquity to the 18th Century* (Archimedes, vol. 37). Cham: Springer.

Bialostocki, Jan. 1988. "Books of Wisdom and Books of Vanity." In Jan Bialostocki (ed.), *The Message of Images: Studies in the History of Art*. Vienna: Irsa.

Biggs, Noah. 1651. *Mataeotechnia medicinae praxeos, The vanity of the craft of physick, or, A new dispensatory wherein is dissected the errors, ignorance, impostures and supinities of the schools ...* London: Calvert.

Bilak, Donna. 2013. "Alchemy and the End Times: Revelations from the Laboratory and Library of John Allin, Puritan Alchemist (1623–1683)." *Ambix*, 60(4): 390–414.

Bilak, Donna. 2014. "The Chymical Cleric: John Allin, Puritan Alchemist in England and America (1623–1683)." Ph.D. thesis, Bard Graduate Center: Decorative Arts, Design History, Material Culture.

Bilak, Donna. 2020. "Chasing Atalanta: Maier, Steganography, and the Secrets of Nature." In Tara Nummedal and Donna Bilak (eds), *Furnace and Fugue: A Digital*

Edition of Michael Maier's Atalanta fugiens (1618) with Scholarly Commentary. Charlottesville: University of Virginia Press.

Biringuccio, Vannoccio. [1540] 1966. *The Pirotechnia*. Cyril Stanley Teach and Martha Teach Gnudi (trans.). Cambridge, MA: MIT Press.

Blair, Ann. 2010. *Too Much to Know: Managing Scholarly Information Before the Modern Age*. New Haven, CT: Yale University Press.

Bleichmar, Daniela, et al. (eds). 2009. *Science and the Spanish and Portugese Empires, 1500–1800*. Stanford, CA: Stanford University Press.

Bostocke, Richard. 1585. *The difference betweene the auncient Phisicke ... and the latter Phisicke*. London: Robert Walley.

Boyle, Robert. 1661. *Certain Physiological Essays and other tracts written at distant times, and on several occasions*. London: Herringman.

Boyle, Robert. 1676. *Experiments, notes, &c. about the mechanical origine or production of divers particular qualities among which is inferred a discourse of the imperfection of the chymist's doctrine of qualities*. London: E. Flesher.

Boyle, Robert. 1680. *Experiments and Notes about the Producibleness of Chymicall Principles*. Published with idem, *The Sceptical Chymist, or, Chymico-Physical Doubts & Paradoxes*. Oxford: Richard Davis.

Boyle, Robert. 1999a. "A Physico-Chymical Essay ... touching ... Saltpetre." In Michael Hunter and Edward B. Davis (eds), *The Works of Robert Boyle*. London: Pickering and Chatto.

Boyle, Robert. 1999b. "Of the Usefulnesse of Natural Philosophy. The First Part. Of its Usefulnesse in reference to *the Mind of Man*." In Michael Hunter and Edward B. Davis (eds), *The Works of Robert Boyle*. London: Pickering and Chatto.

Boyle, Robert. 1999c. "The Origin of Formes and Qualities, (According to the Corpuscular Philosophy,) Illustrated by Considerations and Experiments." In Michael Hunter and Edward B. Davis (eds), *The Works of Robert Boyle*. London: Pickering and Chatto.

Brading, D.A., and Harry E. Cross. 1972. "Colonial Silver Mining: Mexico and Peru." *The Hispanic American Histirical Review*, 52: 545–79.

Brant, Sebastian. 1962. *The Ship of Fools*. Edwin H. Zeydel (trans.). New York: Dover Publications.

Brown, Peter. [1971] 1989. *The World of Late Antiquity:* AD *150–750*. New York: Norton.

Brunschwig, Hieronymus. 1500. *Liber de arte distillandi de simplicibus. Das buch der rechten kunst zü distilieren die eintzigen ding*. Strasbourg: Johann Grüninger.

Brunschwig, Hieronymus. 1512. *Liber de arte distillandi de compositis. Das buch der waren kunst zü distillieren die composita vnn simplicia / vnd dz Buoch thesaurus pauperum / Ein schatz der armen*. Strasbourg: Johann Grüninger.

Brunschwig, Hieronymus. 1527. *The Vertuose Boke of Distyllacyon*. London: Laurens Andrewe.

Brunschwig, Hieronymus. 1973. *The Virtuose boke of Distyllacyon*. Laurence Andrew (trans.). New York: Da Capo Press.

Butterfield, Herbert. 1949. *The Origins of Modern Science, 1300–1800*. London: G. Bell and Sons.

Butters, Suzanne. 2000. "'Una pietra Eppure non Una pietra' Pietre Dure e Botteghe Medicee nella Firenza del Cinquecento." In Franco Franceschi and Gloria Fossi (eds), *Arti Fiorentine: La Grande Storia dell' Artigianato*, vol. 3. Florence: Giunti Gruppo Editoriale.

Calvet, Antoine. 2010. "La théorie *per minima* dans les textes alchimiques des XIVe et XVe siècles." In Miguel Lopez-Perez, Didier Kahn, and Mar Rey Bueno (eds), *Chymia: Science and Nature in Early Modern Europe*. Cambridge: Cambridge Scholars Publishing.
Cardanus, Hieronymus. 1551. *Hieronymi Cardani medici mediolanensis de subtilitate libri XXI* ... Paris: Iacobum Dupuys.
Cerruti, Luigi, Gianmarco Ieluzzi, and Francesca Turco. 2008. "Changing Identity and Public Image: A Sociosemiotic Analysis of Famous Chemical Laboratory Pictures." In *Proceedings of the 6th International Conference on the History of Chemistry*. Louvain: Peeters Publishers.
Chang, Ku Ming (Kevin). 2002. "The Matter of Life: Georg Ernst Stahl and the Reconceptualizations of Matter, Body and Life in Early Modern Europe." Ph.D. thesis, University of Chicago.
Chang, Ku Ming (Kevin). 2015. "Phlogiston and Chemical Principles: The Development and Formulation of Georg Ernst Stahl's Principle of Inflammability." In Karen Hunger Parshall, Michael T. Walton, and Bruce T. Moran (eds), *Bridging Traditions: Alchemy, Chemistry and Paracelsian Practices in the Early Modern Era: Essays in Honor of Allen G. Debus*. Kirksville, MO: Truman State University Press.
Chapman, H. Perry. 1990. *Rembrandt's Self-Portraits: A Study in Seventeenth-Century Identity*. Princeton, NJ: Princeton University Press.
Chaucer, Geoffrey. 1913. *The Works of Geoffrey Chaucer*. Alfred William Pollard (ed.). London: Macmillan and Company Limited.
Christianson, John Robert. 2000. *On Tycho's Island: Tycho Brahe and His Assistants (1570–1601)*. Cambridge: Cambridge University Press.
Clajus, Johannes. 1591. *Altkumistica, das ist, Ein wunderbarliche, seltzsame vnd bewerte Kunst, auß Mist durch seine vilfältige vnd mancherley wirckung Gold zu machen*. Amberg: Michael Forster.
Clericuzio, Antonio. 1993. "From van Helmont to Boyle. A Study of the Transmission of Helmontian Chemical and Medical Theories in Seventeenth-Century England." *The British Journal for the History of Science*, 26(3): 303–34.
Clericuzio, Antonio. 2000. *Elements, Principles, and Corpuscles: A Study of Atomism and Chemistry in the Seventeenth Century*. Dordrecht: Springer.
Clericuzio, Antonio. 2006. "Teaching Chemistry and Chemical Textbooks in France: From Beguin to Lemery." *Science and Education*, 15: 335–55.
Clucas, Stephen. 2007. "Alchemy and Certainty in the Seventeenth Century." In Lawrence M. Principe (ed.), *Chymists and Chymistry: Studies in the History of Alchemy and Early Modern Chemistry*. Sagamore Beach, MA: Watson Publishing.
Cole, Michael. 2002. "The Demonic Arts and the Origin of the Medium." *The Art Bulletin*, 84(4): 621–40.
Conant, James Bryant. 1950. "The Overthrow of the Phlogiston Theory: The Chemical Revolution of 1775–1789." In James Conant (ed.), *Harvard Case Histories in Experimental Science*. Cambridge, MA: Harvard University Press.
Conring, Hermann, 1648. *De hermetica Aegyptiorum vetere et paracelsicorum nova medicina*. Helmstadii: Typis Henningi Mulleri acad. typ. Sumptibus Martini Richteri.
Cook, Harold. 1986. *The Decline of the Old Medical Regime in Stuart London*. Ithaca, NY: Cornell University Press.
Cook, Harold. 1990. "The New Philosophy and Medicine in Seventeenth-Century England." In David Lindberg and Robert Westman (eds), *Reappraisals of the Scientific Revolution*. Cambridge: Cambridge University Press.

Cook, Harold. 2007. *Matters of Exchange: Commerce, Medicine, and Science in the Dutch Golden Age*. New Haven, CT: Yale University Press.

Cooter, Roger, and Stephen Pumfrey. 1994. "Separate Spheres and Public Spaces: Reflections on the History of Science Popularisation and Science in Popular Culture." *History of Science*, 32: 237–67.

Copenhaver, Brian P. 1992. *Hermetica: The Greek Corpus Hermeticum and the Latin Asclepius*. Cambridge: Cambridge University Press.

Corbett, Jane Russell. 2005. "Convention and Change in Seventeenth-Century Depictions of Alchemists." In Jacob Wamberg (ed.), *Art and Alchemy*. Copenhagen: Museum Tusculanum.

Cornejová, Ivana. 1997. "Education in Rudolfine Prague." In Eliška Fucíková (ed.), *Rudolf II and Prague: The Court and the City*. Prague: Prague Castle Administration; London: Thames and Hudson.

Cortese, Isabella. 1561. *I secreti*. Venice: Giovanni Bariletto.

Cressy, David. 2013. *Saltpeter: The Mother of Gunpowder*. Oxford: Oxford University Press.

Crisciani, Chiara, and Michela Pereira. 1998. "Black Death and Golden Remedies. Some Remarks on Alchemy and the Plague." In Agostino Paravicini Bagliani and Francesco Santi (eds), *The Regulation of Evil. Social and Cultural Attitudes to Epidemics in the Late Middle Ages*. Sismel: Edizioni del Galluzzo.

Crosland, Maurice. 2005. "Early Laboratories c.1600–1800 and the Location of Experimental Science." *Annals of Science*, 62(2): 233–53.

Csordas, Thomas J. 1993. "Somatic Modes of Attention." *Cultural Anthropology*, 8: 135–56.

Darmstaedter, Ernst. 1926. *Berg- Probier- und Kunstbüchlein. Münchener Beiträge zur Geschichte und Literatur der Naturwissenschaften und Medizin* (Heft 2/3). Munich: Münchner Drucke.

Daston, Lorraine. 1995. "Curiosity in Early Modern Science." *Word and Image*, 11: 391–404.

Daston, Lorraine, and Katherine Park. 1998. *Wonders and the Order of Nature, 1150–1750*. New York: Zone Books.

Daston, Lorraine, and Elizabeth Lunbeck (eds). 2011. *Histories of Scientific Observation*. Chicago, IL: University of Chicago Press.

Davidson, Jane P. 1979. *David Teniers the Younger*. Boulder, CO: Westview Press.

Davidson, Jane P. 1987. "I Am the Poison Dripping Dragon: Iguanas and Their Symbolism in the Alchemical and Occult Paintings of David Teniers the Younger." *Ambix*, 34(2): 62–80.

De Bertereau, Martine. 1640. *La restitution de Pluton*. Paris: Hervé du Mesnil.

Dear, Peter. 1995. *Discipline and Experience: The Mathematical Way in the Scientific Revolution*. Chicago, IL: Chicago University Press.

Debus, Allen G. 1967. "Fire Analysis and the Elements in the Sixteenth and Seventeenth Centuries." *Annals of Science*, 23: 128–47.

Debus, Allen G. 1977. *The Chemical Philosophy: Paracelsian Science and Medicine in the Sixteenth and Seventeenth Centuries*. New York: Science History Publications.

Debus, Allen G. 1990. "Chemistry and the Universities in the Seventeenth Century." *Estudos Avançados*, 4(10): 173–96.

Debus, Allen G. 1991. *The French Paracelsians: The Chemical Challenge to Medical and Scientific Tradition in Early Modern France*. Cambridge: Cambridge University Press.

Debus, Allen G. 1998. "Paracelsus and the Diffusion of the Chemical Philosophy in Early Modern Europe." In Ole Peter Grell (ed.), *Paracelsus, the Man and His Reputation, His Ideas and Their Transformation*. Leiden/Boston, MA: Brill.

Debus, Allen G. 2001. *Chemistry and Medical Debate: van Helmont to Boerhaave*. Canton, MA: Science History Publications.

De Chauliac, Guy. 1580. *La Grand Chirurgie*. Laurent Joubert (trans.). Lyon: Estienne Michel.

De Jong, H.M.E. 1969. *Michael Maier's "Atalanta fugiens": Sources of an Alchemical Book of Emblems*. Leiden: E.J. Brill.

De Vivo, Filippo. 2007. "Pharmacies as Centres of Communication in Early Modern Venice." *Renaissance Studies*, 21: 505–21.

DeVun, Leah. 2009. *Prophecy, Alchemy, and the End of Time: John of Rupescissa in the Late Middle Ages*. New York: Columbia University Press.

Dorn, Gerhard. 1567. *Clavis totius philosophiæ chymisticæ*. Lyons: apud héritiers Jacques Giunta.

Du Chesne, Joseph. 1603. *Liber de priscorum philosophorum verae medicinae materia, …* Saint-Gervais: Apud Haeredes Eustathii Vignon.

Du Chesne, Joseph. 1604. *Ad veritatem hermeticae medicinae ex Hippocratis veterumque decretis ac therapeusi*. Paris: Abraham Saugrain.

Dupré, S. 2014a. "The Value of Glass and the Translation of Artisanal Knowledge in Early Modern Antwerp." In C. Göttler, B. Ramakers, and J. Woodall (eds), *Trading Values in Early Modern Antwerp*. Leiden: Brill.

Dupré, Sven (ed.). 2014b. *Laboratories of Art: Alchemy and Art Technology from Antiquity to the 18th Century*. Cham: Springer.

Dupré, Sven, Dedo von Kerssenbrock-Krosigk, and Beat Wismer (eds). 2014. *Art and Alchemy: The Mystery of Transformation*. Munich: Hirmer; Düsseldorf: Stiftung Museum Kunstpalast.

Eamon, William. 1980. "New Light on Robert Boyle and the Discovery of Colour Indicators." *Ambix*, 27: 204–9.

Eamon, William. 1994. *Science and the Secrets of Nature: Books of Secrets in Medieval and Early Modern Culture*. Princeton, NJ: Princeton University Press.

Eamon, William. 2000. "Alchemy in Popular Culture: Leonardo Fioravanti and the Search for the Philosopher's Stone." *Early Science and Medicine*, 5: 196–213.

Eamon, William. 2003. "Pharmaceutical Self-fashioning or How to Get Rich and Famous in the Renaissance Medical Marketplace." *Pharmacy in History*, 45: 123–9.

Eamon, William. 2006. "Markets, Piazzas, and Villages." In K. Park and L. Daston (eds), *The Cambridge History of Science*, vol. 3. Cambridge: Cambridge University Press.

Eamon, William. 2010. *The Professor of Secrets: Mystery, Medicine, and Alchemy in Renaissance Italy*. Washington, DC: Smithsonian.

Eamon, William. 2016. "The Scientific Education of a Renaissance Prince: Archduke Rudolf at the Spanish Court." In Ivo Purš and Vladimír Karpenko (eds), *Alchemy and Rudolf II. Exploring the Secrets of Nature in Central Europe in the 16th and 17th Centuries*. Prague: Artefactum.

Eamon, William. 2018. "Corn, Cochineal, and Quina: The Zilsel Thesis in an Iberian Setting." *Centaurus*, 60(3): 141–58.

Eisenstein, Elizabeth. 1979. *The Printing Press as an Agent of Change: Communications and Cultural Transformations in Early-Modern Europe*. Cambridge: Cambridge University Press.

Erastus, Thomas. [1571]–3. *Disputationes de medicina nova Philippi Paracelsi*. Basel: Peter Perna.

Ercker, Lazarus. 1574. *Beschreibung allerfürnemisten mineralischen Ertzt unnd Berckwercksarten, wie dieselbigen, unnd eine jede insonderheit, der Natur und Eigenschafft nach, auff alle Metale Probirt ... in fünff Bücher verfast*. Prague: Georg Schwartz.

Ercker, Lazarus. 1951. *Treatise on Ores and Assaying*. Anneliese Grünhaldt Sisco and Cyril Stanley Smith (trans.). Chicago, IL: University of Chicago Press.

Evans, R.J.W. 1973. *Rudolf II and His World: A Study in Intellectual History*. Oxford: Oxford University Press.

Evans, R.J.W. 1979. *The Making of the Habsburg Monarchy 1550–1700*. Oxford: Oxford University Press.

Evans, R.J.W., and Alexander Marr (eds). 2006. *Curiosity and Wonder from the Renaissance to the Enlightenment*. Farnham: Ashgate.

Evelyn, John. 1776. *Silva: Or, a Discourse of Forest-Trees and the Propagation of Timber in His Majesty's Dominions*. York: A. Ward.

Feldman, Richard. 2003. "Epistemology." In Tom L. Beauchamp (ed.), *Foundations of Philosophy*, Upper Saddle River, N.J.: Prentice Hall.

Ferguson, John K. 1959. *Bibliographical Notes on Histories of Inventions and Books of Secrets*. London, Holland Press.

Ferrario, Gabriele. 2009. "Understanding the Language of Alchemy: The Medieval Arabic Alchemical Lexicon in Berlin, Staatsbibliothek, Ms Sprenger 1908*." *Digital Proceedings of the Lawrence J. Schoenberg Symposium on Manuscript Studies in the Digital Age*, 1(1): Article 2. Available online: http://repository.upenn.edu/ljsproceedings/vol1/iss1/2.

Ferrario, Gabriele. 2010. "The Jews and Alchemy: Notes for a Problematic Approach." In M. López Pérez, D. Kahn, and M. Rey Bueno (eds), *Chymia. Science and Nature in Medieval and Early Modern Europe*. Newcastle upon Tyne: Cambridge Scholars.

Fichman, Martin. 1971. "French Stahlism and Chemical Studies of Air, 1750–1770." *Ambix*, 18: 94–122.

Figala, Karin, and Ulrich Newmann. 1990. "Michael Maier (1569–1622): New Bio-Bibliographical Material." In Z.R.M.W. Von Martels (ed.), *Alchemy Revisited: Proceedings of an international congress at the University of Groningen, 17–19 April 1989*. Leiden: Brill.

Findlen, Paula. 1997. "Cabinets, Collecting and Natural Philosophy." In Eliška Fučíková (ed.), *Rudolf II and Prague: The Court and the City*. Prague: Prague Castle Administration; London: Thames and Hudson.

Fioravanti, Leonardo. 1579. *A ioyfull iewell Contayning aswell such excellent orders, preseruatiues and precious practises for the plague, as also such meruelous medcins for diuers maladies, as hitherto haue not beene published in the English tung*. London: Wright.

Fissell, Mary. 1992. "Readers, Texts, and Contexts: Vernacular Medical Works in Early Modern England." In Roy Porter (ed.), *The Popularization of Medicine, 1650–1850*. London: Routledge.

Fissell, Mary. 2007. "The Marketplace of Print." In P. Wallis and M. Jenner (eds), *Medicine and the Market in England and Its Colonies, c.1450–c.1850*. Basingstoke: Palgrave Macmillan.

Forbes, R.J. 1948. *Short History of Distillation*. Leiden: Brill.

Forbes, Robert. 1970. *A Short History of the Art of Distillation: From the Beginnings Up to the Death of Cellier Blumenthal*. Leiden: Brill.
Forbes Sieveking, A. 1923–4. "Evelyn's 'Circle of Mechanical Trades'." *The Newcomen Society for the Study of the History of Engineering and Technology Transactions*, 4: 40–7.
Fors, Hjalmar, Lawrence Principe, and Otto Sibum. 2016. "From the Library to the Laboratory and Back Again: Experiment as a Tool for Historians of Science." *Ambix*, 63: 85–97.
Forshaw, Peter. 2010. "Oratorium – Auditorium – Laboratorium: Early Modern Improvisations on Cabala, Music, and Alchemy." *Aries*, 10(2): 169–95.
Fowden, Garth. 1986. *The Egyptian Hermes: A Historical Approach to the Late Pagan Mind*. Cambridge: Cambridge University Press.
Franits, Wayne E. 1993. *Paragons of Virtue: Women and Domesticity in Seventeenth-Century Dutch Art*. Cambridge: Cambridge University Press.
Freedberg, David. 2002. *The Eye of the Lynx: Galileo, His Friends, and the Beginnings of Modern Natural History*. Chicago, IL: University of Chicago Press.
French, John. 1651. *The Art of Distillation*. London: Cotes.
Fucíková, Eliška. 1997. "Prague Castle Under Rudolf II: His Predecessors and Successors 1530–1648." In Eliška Fucíková (ed.), *Rudolf II and Prague: The Court and the City*. Prague: Prague Castle Administration; London: Thames and Hudson.
Galison, Peter. 1995. "Context and Constraints." In Jed Z. Buchwald (ed.), *Scientific Practice: Theories and Stories of Doing Physics*. Chicago, IL: University of Chicago Press.
Galison, Peter. 1997. "The Trading Zone: Coordinating Action and Belief." In Peter Galison (ed.), *Image and Logic: A Material Culture of Microphysics*. Chicago, IL: University of Chicago Press.
Galison, Peter. 2010. "Trading with the Enemy." In Michael E. Gorman (ed.), *Trading Zones and Interactional Expertise: Creating New Kinds of Collaboration*. Cambridge, MA: MIT Press.
Garber, Margaret D. 2005. "Chymical Wonders of Light: J. Marcus Marci's Seventeenth-Century Bohemian Optics." *Early Science and Medicine*, 10(4): 478–509.
Garber, Margaret D. 2007. "Transitioning from Transubstantiation to Transmutation: Catholic Anxieties over Chemical Matter Theory at the University of Prague." In Lawrence M. Principe (ed.), *Chymists and Chymistry: Studies in the History of Alchemy and Early Modern Chemistry*. Sagamore Beach, MA: Watson Publishing.
Garber, Margaret D. 2015. "Chymical Curiosities and Trusted Testimonials in the Journal of the Leopoldina Academy of Curiosi." In Karen Hunger Parshall, Michael T. Walton, and Bruce T. Moran (eds), *Bridging Traditions: Alchemy, Chemistry, and Paracelsian Practices in the Early Modern Era*. Kirksville, MO: Truman State University Press.
García-Ballester, Luis. 2002. *Galen and Galenism: Theory and Medical Practice from Antiquity to the European Renaissance*. Burlington, VT: Ashgate/Variorum.
Garzoni, Tomaso. [1585] 1996. *La piazza universale di tutte le professioni del mondo*. Paolo Cherchi (ed.). Turin: Einaudi.
Gaukroger, Stephen. 2001. *Francis Bacon and the Transformation of Early-Modern Philosophy*. Cambridge: Cambridge University Press.
Gentilcore, David. 1998. *Healers and Healing in Early Modern Italy*. Manchester: Manchester University Press.

Geertz, Clifford. 1983. "Local Knowledge: Fact and Law in Comparative Perspective." In Clifford Geertz (ed.), *Local Knowledge: Further Essays in Interpretive Anthropology*. New York: Basic Books.

Gentilcore, David. 1995. "Charlatans, Mountebanks and Other Similar People: The Regulation and Role of Itinerant Practitioners in Early Modern Italy." *Social History*, 20(3): 297–314.

Gessner, Conrad. 1552. *Thesaurus Euonymi Philiatri de remediis secretis, liber physicus, medicus et partim etiam chymicus, & oeconomicus ...* Zurich: Gessner.

Gessner, Conrad. 1555. *Ein kostlicher theürer Schatz Euonymi Philiatri*. Zurich: Gessner.

Gessner, Conrad. 1576. *The newe iewell of health wherein is contayned the most excellent secretes of phisicke and philosophie*. London: Denham.

Glacken, Clarence. 1967. *Traces on the Rhodian Shore. Nature and Culture in Western Thought from Ancient Times to the End of the Eighteenth Century*. Berkeley: University of California Press.

Glauber, Johann Rudolf. 1651. *A description of new philosophical furnaces: or, A new art of distilling, divided into five parts*. London: s.n.

Glisson, Francis. 1654. *Anatomia Hepatis*. London: Octavius Pulleyn.

Glorez, Andreas. 1699. *Der vollständigen Hauß- und Land-Bibliothec*. Regensburg: Quirin Heil.

Glorez, Andreas (ed.). 1701. *Continuation der vollständigen Hauß- und Land-Bibliothec*, vol. 3. Regensburg: Quirin Heil.

Golinski, Jan V. 1992. *Science as Public Culture: Chemistry and Enlightenment in Britain, 1760–1820*. Cambridge: Cambridge University Press.

Goodman, David C. 1988. *Power and Penury: Government, Technology, and Science in Philip II's Spain*. Cambridge: Cambridge University Press.

Gough, J.B. 1988. "Lavoisier and the Fulfillment of the Stahlian Revolution." *Osiris*, 2nd Series(4): 15–33.

Gratarolo, Guglielmo (ed.). 1572. *Auriferæ artis, quam chemiam vocant aintiquissimi authores, sive, Turba philosophorum*. Basel: Peter Perna.

Green, Tamara M. 1992. *The City of the Moon God: Religious Traditions of Harran*. Leiden and New York: Brill.

Greenberg, Arthur. 2007. *From Alchemy to Chemistry in Picture and Story*. Hoboken, NJ: Wiley-Interscience.

Greenfield, Amy Butler. 2005. *A Perfect Red: Empire, Espionage, and the Quest for the Color of Desire*. New York: Harper.

Guerrini, Anita. 1994. "Chemistry Teaching at Oxford and Cambridge, circa 1700." In Piyo Rattansi and Antonio Clericuzio (ed.), *Alchemy and Chemistry in the Sixteenth and Seventeenth Centuries*. Dordrecht: Kluwer Academic Publishers.

Gussman, Neil. 2015. "Books of Secrets: Writing & Reading Alchemy." *Chemistry International*, 37(2): 4–6.

Hacking, Ian. 1986. "Making Up People." In Thomas C. Heller, David E. Wellbery, and Morton Sosna (eds.), *Reconstructing Individualism: Autonomy, Individuality, and the Self in Western Thought*, 222–36. Stanford: Stanford University Press.

Hale, John R. 1994. *The Civilization of Europe in the Renaissance*. New York: Atheneum.

Hall, Bert S. 1997. *Weapons and Warfare in Renaissance Europe: Gunpowder, Technology, and Tactics*. Baltimore, MD: Johns Hopkins University Press.

Hannaway, Owen. 1975. *The Chemist and the Word: The Didactic Origin of Chemistry*. Baltimore, MD: Johns Hopkins University Press.

Hannaway, Owen. 1986. "Laboratory Design and the Aim of Science: Andreas Libavius versus Tycho Brahe." *Isis*, 77(4): 584–610.

Harkness, Deborah. 2007. *The Jewel House: Elizabethan London and the Scientific Revolution*. New Haven, CT: Yale University Press.

Harrington, Joel F. 2013. *The Faithful Executioner: Life and Death, Honor and Shame in the Turbulent Sixteenth Century*. New York: Farrar, Straus and Giroux.

Hartlib, Samuel, et al. 2013. *The Hartlib Papers*. Available online: https://www.dhi.ac.uk/hartlib/.

Heinrichs, Erik Anton. 2012. "The Plague Cures of Caspar Kegler: Print, Alchemy, and Medical Marketing in Sixteenth-Century Germany." *Sixteenth Century Journal*, 63: 417–40.

Hedesan, Georgiana D. 2014. "Paracelsian Medicine and Theory of Generation in 'Exterior homo', a Manuscript Probably Authored by Jan Baptist Van Helmont (1579–1644)." *Medical History*, 58(3): 375–96.

Helvetius, Johann Friedrich. 1670. *The Golden Calf, Which the World Adores and Desires*. London: John Starkey.

Hester, John. 1575. *The true and perfect order to distill oyles out of al maner of spices seedes, rootes, and gummes*. London: s.n.

Hester, John. 1585. *These oiles, vvaters, extractions, or essence[s,] saltes, and other compositions; are at Paules wharfe ready made to be solde, by Iohn Hester, practisioner in the arte of distillation*. London: s.n.

Hirai, Hiro. 2005. *Le concept de semence dans les théories de la matière à la Renaissance de Marsile Ficin à Pierre Gassendi*. Turnhout: Brepols.

Hirai, Hiro. 2007. "Kircher's Chymical Interpretation of the Creation and Spontaneous Generation." In Lawrence M. Principe (ed.), *Chymists and Chymistry: Studies in the History of Alchemy and Early Modern Chemistry*. Sagamore Beach, MA: Watson Publishing.

Hirsch, Rudolf. 1967. *Printing, Selling, and Reading 1450–1550*. Wiesbaden: Otto Harrassowitz.

Hofmann, Caspar. 1726. "Oratio de barbarie immenente." In Joachimus Negelein (ed.), *Ulysses literarius sive oratio de singularibus et novis quibusdam ex orbe literato* ... Norimbergae: ex officina Wolf. Maur. Endteriana.

Hooykaas, Reijer. 1935. "Die Elementenlehre des Paracelsus." *Janus*, 39: 175–87.

Hooykaas, Reijer. 1937. "Die Elementenlehre der Iatrochemiker." *Janus*, 41: 1–28.

Hooykaas, Reijer. 1949. "The Experimental Origin of Chemical Atomic and Molecular Theory before Boyle." *Chymia*, 2: 65–80.

Hooykaas, Reijer. 1983. *The Concept of Element: Its Historical-Philosophical Development*. Utrecht: Drukkerij fa. Schotanus & Jens.

Houbraken, Arnold. 1976. *De groote schouburgh der Nederlantsche konstschilders en schilderessen*. Amsterdam: B.M. Israël.

Hufton, Olwin. 1974. *The Poor of Eighteenth Century France 1750–1789*. Oxford: Clarendon Press.

Hunter, Michael. 2009. *Boyle: Between God and Science*. New Haven, CT: Yale University Press.

Jacob, Margaret C. 2006. *Strangers Nowhere in the World: The Rise of Cosmopolitanism in Early Modern Europe*. Philadelphia: University of Pennsylvania Press.

Jacob, Margaret C. 2008. "The Cosmopolitan as a Lived Category." *Daedalus*, 137: 18–25.

Jenner, Mark, and Patrick Wallis (eds). 2007. *Medicine and the Market in England and Its Colonies, c.1450–c.1850*. Basingstoke: Palgrave Macmillan.

Jodziewicz, Thomas W. 1988. "A Stranger in the Land: Gershom Bulkeley of Connecticut." *Transactions of the American Philosophical Society, New Series*, 78(2): 1–106.

Johns, Adrian. 1998. *The Nature of the Book: Print and Knowledge in the Making*. Chicago, IL: University of Chicago Press.

Joly, Bernard. 1996. "L'alkahest, dissolvant universel, ou quand la thèorie rend pensible une pratique impossible." *Revue d'histoire des sciences*, 49: 308–30.

Joly, Bernard. 2001. "La thèorie des cinq éléments d'Ètienne de Clave dans la *Nouvelle Lumière Philosophique. Dossier Étienne de Clave*." *Corpus: Revue de philosophie*, 39: 9–44.

Joly, Bernard. 2011. *Descartes et la chimie*. Paris: Vrin.

Jonson, Ben. 1612. *The Alchemist*. London: Walter Burre and John Stepneth.

Jonson, Ben. 1903. *The Alchemist*. Charles Montgomory Hathaway (ed.). London: H. Hart/De La More.

Kahn, Didier. 2007a. *Alchimie et Paracelsisme en France à la fin de la Renaissance (1567–1625)*. Geneva: Librairie Droz.

Kahn, Didier. 2007b. "King Henry IV, Alchemy, and Paracelsianism in France (1589–1610)." In Lawrence M. Principe (ed.), *Chymists and Chymistry: Studies in the History of Alchemy and Early Modern Chemistry*. Sagamore Beach, MA: Watson Publishing.

Kahn, Didier. 2014. "The Significance of Transmutation in Early Modern Alchemy: The Case of Thurneysser's Half-Gold Nail." In Marco Beretta and Maria Conforti (eds), *Fakes!? Hoaxes, Counterfeits and Deception in Early Modern Science*. Sagamore Beach, MA: Science History Publications/Watson Publishing International.

Kassell, Lauren. 2011. "Secrets Revealed: Alchemical Books in Early Modern England." *History of Science*, 49: 63–87.

Kaufmann, Thomas DaCosta. 1988. *The School of Prague: Painting at the Court of Rudolf II*. Chicago, IL: Chicago University Press.

Kaufmann, Thomas DaCosta. 1997. "Kunst und Alchemie." In Heiner Borggrefe, Vera Lüpkes, and Hans Ottomeyer (eds), *Moritz der Gelehrte – Ein Renaissancefürst in Europa*. Eruasberg: Edition Minerva.

Keller, Vera. 2012. "Mining Tacitus: Secrets of Empire, Nature and Art in the Reason of State." *The British Journal for the History of Science*, 45: 189–212.

Keller, Vera. 2015. *Knowledge and the Public Interest, 1575–1725*. Cambridge: Cambridge University Press.

Kellman, Jordan. 2010. "Nature, Networks, and Expert Testimony in the Colonial Atlantic: The Case of Cochineal." *Atlantic Studies*, 7: 373–95.

Kenny, Neil. 1998. *Curiosity in Early Modern Europe: Word Histories* (Wolfenbütteler Forschungen, 81). Wiesbaden: Harrassowitz.

Kenny, Neil. 2004. *The Uses of Curiosity in Early Modern France and Germany*. Oxford: Oxford University Press.

Kenyon, Ralph. 1994. "On the Use of Quotation Marks." *Etc.: A Review of General Semantics*, 51: 47–50.

Kerzenmacher, Peter. 1531. *Rechter Gebrauch d'Alchimei*. Frankfurt am Main: Christian Egenolff the Elder.

Kerzenmacher, Peter. 1534. *Alchimi und Bergwerck*. Strasbourg: Jakob Cammerlander.

Kerzenmacher, Peter. 1538. *Alchimia: Wie man[n] alle Farben, Wasser, olea, salia und alumina, darmit mann alle corpora, spiritus und calces preparirt, sublimirt und fixirt, machen sol. Und wie mann dise ding nutze, auff das sol und luna werden mög. Auch vom soluiren unnd Schaidung aller Metal*. Strasbourg: Jakob Cammerlander.

Kettering, Alison M. 2007. "Men at Work in Dutch Art, or Keeping One's Nose to the Grindstone." *The Art Bulletin*, 89(4): 694–714.
Khunrath, Heinrich. 1595. *Amphitheatrum sapientiae aeternae*. Hamburg: s.n.
Kieffer, Fanny. 2014. "The Laboratories of Art and Alchemy at the Uffizi Gallery in Renaissance Florence." In Sven Dupré (ed.), *Laboratories of Art, Alchemy, and Art Technology from Antiquity to the Eighteenth Century* (Archimedes, vol. 37). Cham: Springer.
Kim, Mi Gyung. 2003. *Affinity, That Elusive Dream: A Genealogy of the Chemical Revolution*. Cambridge, MA:MIT Press.
Kirby, Jo, Susie Nash, and Joanna Cannon. 2010. *Trade in Artists' Materials: Markets and Commerce in Europe to 1700*. London: Archetype Publications.
Klein, Joel A. 2014a. "Chymical Medicine, Corpuscularism, and Controversy: A Study of Daniel Sennert's Works and Letters." Ph.D. thesis, Indiana University, Bloomington.
Klein, Joel A. 2014b. "Corporeal Elements and Principles in the Learned German Chymical Tradition." *Ambix*, 61(4): 345–65.
Klein, Joel A. 2015. "Daniel Sennert, The Philosophical Hen, and The Epistolary Quest for a (Nearly-)Universal Medicine." *Ambix*, 62(1): 29–49.
Klein, Joel A. 2016a. "Daniel Sennert and the Chymico-atomical Reform of Medicine." In Ole Peter Grell and Andrew Cunningham (eds), *Medicine, Natural Philosophy and Religion in Post-Reformation Scandanavia*. London/New York: Routledge.
Klein, Joel A. 2016b. "Alchemical Histories, Chymical Education, and Chymical Medicine in Sixteenth- and Seventeenth-Century Wittenberg." In Harald Meller, Alfred Reichenberger, Christian-Heinrich Wunderlich (eds), *Alchemie und Wissenschaft des 16. Jahrhunderts. Fallstudien aus Wittenberg und vergleichbare Befunde*. Halle: Landesamt für Denkmalpflege und Archäologie Sachsen-Anhalt, Landesmuseum für Vorgeschichte.
Klein, Joel A., and Evan Ragland. 2014. "Analysis and Synthesis in Medieval and Early-Modern Europe." *Ambix*, 61: 319–26.
Klein, Ursula. 2008. "The Laboratory Challenge: Some Revisions of the Standard View of Early. Modern Experimentation." *Isis*, 99: 769–82.
Klein, Ursula, and Emma Spary (eds). 2010. *Materials and Expertise in Early Modern Europe: Between Market and Laboratory*. Chicago, IL: University of Chicago Press.
Knight, David. 1992. *Ideas in Chemistry: A History of the Science*. London: Athlone Press.
Knox, Kevin C. 2005. "'The Deplorable Frenzy': The Slow Legitimisation of Chemical Practice at Cambridge University." In Mary Archer and Christopher Haley (eds), *The 1702 Chair of Chemistry at Cambridege: Transformation and Change*. Cambridge: Cambridge University Press.
Kopp, Hermann. 1886. *Die Alchemie bis zum letzten Viertel des 18. Jahrhunderts*. Vol. 1 of his *Die Alchemie in älterer und neuerer Zeit: Ein Beitrag zur Culturgeschichte*. Heidelberg: Carl Winters Universitätsbuchhandlung.
Krafftheim, Crato von. 1582. "Epistola: Praestantissimo D. Iosepho Scaligero, summi viri Iulii Caesaris F. Iohannes Crato a Crafftheim, Archiatrus Caesareus, S. D." In *Iulii Caesaris Scaligeri exotericarum exercitationum lib. XV de subtilltate ad Hieronymum Cardanum*. Francofurti: apud haeredes Andr. Wecheli.
Kristeller, Paul Oskar. 1979. *Renaissance Thought and Its Sources*. Michael Mooney (ed.). New York: Columbia University Press.

Kuhn, Thomas S. [1962] 1970. *The Structure of Scientific Revolutions*. Chicago, IL: University of Chicago Press.
Le Pelletier, Jean. 1706. *L'Alkaest,ou le dissolvant universel de Van Helmont*. Rouen: s.n.
Lee, R.L. 1951. "American Cochineal in European Commerce, 1526–1625." *Journal of Modern History*, 23: 205–24.
LeFèvre, Nicaise. 1662. *A Compendious Body of Chymistry*. London: Davies and Sadler.
Leitão, Henrique, and Antonio Sánchez. 2017. "Zilsel's Thesis, Maritime Culture, and Iberian Science in Early Modern Europe." *Journal of the History of Ideas*, 78: 191–210.
Lenke, N., N. Roudet, and H. Tilton. 2014. "Michael Maier – Nine Newly Discovered Letters." *Ambix*, 61: 1–47.
Leong, Elaine. 2008. "Making Medicines in the Early Modern Household." *Bulletin of the History of Medicine*, 82: 145–68.
Leong, Elaine. 2013. "Collecting Knowledge for the Family: Recipes, Gender and Practical Knowledge in the Early Modern English Household." *Centaurus*, 55: 81–103.
Levens, Peter. 1582. *A right profitable booke for all disseases Called The pathway to health: wherein are to be found most excellent and approued medicines of great vertue, as also notable potions and drinkes, and for the destillinge of diuers pretious waters*. London: White.
Linden, Stanton J. 2003. *The Alchemy Reader: From Hermes Trismegistus to Isaac Newton*. New York: Cambridge University Press.
Lindsay, Jack. 1970. *The Origins of Alchemy in Graeco-Roman Egypt*. London: Frederick Muller.
Lis, Catharina, and Hugo Soly. 2012. *Worthy Efforts: Attitudes to Work and Workers in Pre-Industrial Europe*. London: Brill.
Long, Pamela. 2011. *Artisan/Practitioners and the Rise of the New Sciences, 1400–1600*. Corvallis: Oregon State University Press.
Lonicer, Adam. 1564. *Kreuterbuoch Künstliche Conterfeytunge der Baeume / Stauden / Hecken / Kreuter / Getreyde / Gewürtze ...* Frankfurt: Egenolff heirs.
López Piñero, José María. 1977. *El "Dialogus" (1589) del paracelsista Llorenç Coçar y la cátedra de medicamentos químicos de la Universidad de Valencia (1591)*. Valencia: Cátedra e Instituto de Historia de la Medicina.
López Terrada, María Luz. 2005. "Llorenç Coçar: protomédico de Felipe II y médico paracelsista en la Valencia del siglo XVI." *Cronos: Cuadernos valencianos de historia de la medicina y de la ciencia*, 8: 31–66.
Machiavelli, Niccolo. 2003. *The Art of War*. Christopher Lynch (trans.). Chicago, IL: University of Chicago Press.
Maier, Michael. 1617. *Examen fucorum pseudo-chymicorum detectorum et in gratiam veritatis amantium succinte refutatorum*. Frankfurt am Main: The Heirs of Theodor de Bry the elder.
Maier, Michael. 1618a. *Atalanta fugiens: hoc est, emblemata nova de secretis naturae, chymica*. Oppenheim: For Johann Theodor de Bry.
Maier, Michael. 1618b. *Tripus aureus, hoc est, tres tractatus chymici selectissimi*. Frankfurt: Paul Jacob for Lucas Jennis.
Mandosio, Jean-Marc. 1990–1. "La place de l'alchimie dans les classifications des sciences et des arts à la Renaissance." *Chrysopœia*, 4: 199–282.

Mandosio, Jean-Marc. 1993. "L'alchimie dans les classifications des sciences et des arts à la Renaissance." In Jean-Claude Margolin and Sylvain Matton (eds), *Alchimie et philosophie à la Renaissance: actes du Colloque International de Tours, 4–7 décembre 1991*. Paris: Vrin.

Marci, Johannes Marcus. 1662. *Philosophia vetus restituta*. Frankfurt and Leipzig: Christian Weidmann.

Markey, Lia. 2012. "Stradano's Allegorical Invention of the Americas in Late Sixteenth-Century Florence." *Renaissance Quarterly*, 65(2): 385–442.

Markham, Gervase. [1651] 1675. *The English House-Wife*. London: George Sawbridge.

Marshall, Peter. 2009. *The Magic Circle of Rudolf II: Alchemy and Astrology in Renaissance Prague*. New York: Bloomsbury.

Martelli, Matteo. 2011. "Greek Alchemists at Work: 'Alchemical Laboratory' in the Greco-Roman Egypt." *Nuncius*, 26: 271–311.

Martin, Julian. 1992. *Francis Bacon, the State and the Reform of Natural Philosophy*. Cambridge, MA: Cambridge University Press.

Martin, Wayne M. 2011. "Bubbles and Skulls: The Phenomenological Structure of Self-Consciousness." In Hubert L. Dreyfus and Mark A. Wrathall (eds), *A Companion to Phenomenology and Existentialism*. Oxford: Wiley-Blackwell.

Martinón-Torres, Marcus. 2007. "The Tools of the Chymist: Archaeological and Scientific Analyses of Early Modern Laboratories." In Lawrence M. Principe (ed.), *Chymists and Chymistry: Studies in the History of Alchemy and Early Modern Chemistry*. Sagamore Beach, MA: Watson Publishing.

Marx, Karl. 1988. "The Power of Money in Bourgeois Society." In Martin Milligan (trans.), *Economic and Philosophic Manuscripts of 1844*. Buffalo, NY: Prometheus Books.

Matthew, Louisa. 2011. "The Pigment Trade in Europe during the Sixteenth Century." In Gerhard Wolf, Joseph Connors, and Louis Alexander Waldman (eds), *Colors between Two Worlds: The Florentine Codex of Bernardino de Sahagún* (Villa I Tatti Studies, 28). Florence: Kunsthistorisches Institut in Florenz.

Mattioli, Pier Andrea. 1561. *Epistolarum medicinalium liber quinque*. Prague: In officina Georgij Melantrichij ab Auentino ad instantiam Vincentij Valgrisij.

Matus, Zachary A. 2017. *Franciscans and the Elixir of Life: Religion and Science in the Later Middle Ages*. Philadelphia: University of Pennsylvania Press.

McEvoy, John G. 2000. "In Search of the Chemical Revolution: Interpretive Strategies in the History of Chemistry." *Foundations of Chemistry*, 2: 47–73.

Meinel, Christoph. 1988. "Early Seventeenth-Century Atomism: Theory, Epistemology, and the Insufficiency of Experiment." *Isis*, 79(1): 68–103.

Michael, Emily. 1997. "Daniel Sennert on Matter and Form: At the Juncture of the Old and the New." *Early Science and Medicine*, 2: 272–99.

Miskimin, Harry A. 1977. *The Economy of Later Renaissance Europe*. Cambridge: Cambridge University Press.

Mocchi, Guiliana. 1990. *Idea, mente, specie: Platonizmo e scienza in Johannes Marcus Marci 1595–1667*. Soveria Manelli (Cz.): Rubbettino Editore.

Mola, Luca. 2000. *The Silk Industry of Renaissance Venice*. Baltimore, MD: Johns Hopkins University Press.

Moran, Bruce T. 1991a. *Chemical Pharmacy Enters the University: Johannes Harmann and the Didactic Care of Chymiatria in the Early Seventeenth Century*. Madison, WI: American Institute of the History of Pharmacy.

Moran, Bruce T. (ed.). 1991b. *Patronage and Institutions: Science, Technology, and Medicine at the European Court, 1500–1750*. Woodbridge: Boydell Press.

Moran, Bruce T. 1991c. *The Alchemical World of the German Court: Occult Philosophy and Chemical Medicine in the Circle of Moritz of Hessen (1572–1632)*. Stuttgart: Franz Steiner Verlag.

Moran, Bruce T. 1996. "A Survey of Chemical Medicine in the 17th Century: Spanning Court, Classroom, and Cultures." *Pharmacy in History*, 38(3): 121–33.

Moran, Bruce T. 2005. *Distilling Knowledge: Alchemy, Chemistry, and the Scientific Revolution*. Cambridge, MA: Harvard University Press.

Moran, Bruce T. 2007. *Andreas Libavius and the Transformation of Alchemy: Separating Chemical Cultures with Polemical Fire*. Sagamore Beach, MA: Science History Publications.

Moran, Bruce T. 2014. "Eloquence in the Marketplace: Erudition and Pragmatic Humanism in the Restoration of Chymia." *Osiris*, 29: 49–62.

Moran, Bruce T. 2015. "Andreas Libavius and the Art of Chymia: Words, Works, Precepts, and Social Practices." In Karen Hunger Parshall et al. (eds), *Bridging Traditions: Alchemy, Chemistry, and Paracelsian Practices in the Early Modern Era*. Kirksville, MO: Truman State University Press.

Moran, Bruce T. 2019. *Paracelsus: An Alchemical Life*. London: Reaktion Books.

Morgenstern, Philipp (ed.). 1613. *Turba philosophorum, das ist, Das Buch von der güldenen Kunst neben andern Authoribus*. Basel: Ludwig König and Johann Schröter.

Morris, Peter John Turnbull. 2015. *The Matter Factory: A History of the Chemistry Laboratory*. London: Reaktion Books, in Association with Science Museum.

Mosley, Adam. 2007. *Bearing the Heavens: Tycho Brahe and the Astronomical Community of the Late Sixteenth Century*. Cambridge: Cambridge University Press.

Mukherjee, Ayesha. 2011. "The Secrets of Sir Hugh Platt." In Elaine Leong and Alisha Rankin (eds), *Secrets and Knowledge in Medicine and Science, 1500–1800*. Farnham: Ashgate Publishing, Ltd.

Mukherjee, Ayesha. 2014. *Penury into Plenty: Dearth and the Making of Knowledge in Early Modern England*. London: Routledge.

Müller-Grzenda, Astrid. 1996. *Pflanzenwässer und gebrannter Wein als Arzneimittel zu Beginn der Neuzeit*. Stuttgart: Deutscher Apotheker Verlag.

Multhauf, Robert. 1954. "Medical Chemistry and the Paracelsians." *Bulletin of the History of Medicine*, 28: 101–26.

Multhauf, Robert. 1956. "The Significance of Distillation in Renaissance Medical Chemistry." *Bulletin of the History of Medicine*, 30(4): 329–46.

Multhauf, Robert. 1967. *The Origins of Chemistry*. New York: Franklin Watts.

Musgrave, Peter. 1999. *The Early Modern Economy*. New York: St Martin's Press.

Neri, Antonio. [1612] 1662. *The Art of Glass*. London: Octavian Pulleyn.

Newman, William. 1989. "Technology and Alchemical Debate in the Late Middle Ages." *Isis*, 80(3): 423–45.

Newman, William. 1991. *The Summa perfectionis of Pseudo-Geber: A Critical Edition, Translation, and Study*. Leiden: Brill.

Newman, William. 1994. *Gehennical Fire: The Lives of George Starkey, an American Alchemist in the Scientific Revolution*. Cambridge, MA: Harvard University Press.

Newman, William. 2004. *Promethean Ambitions: Alchemy and the Quest to Perfect Nature*. Chicago, IL and London: University of Chicago Press.

Newman, William. 2006a. *Atoms and Alchemy*. Chicago, IL: University of Chicago Press.
Newman, William. 2006b. "From Alchemy to 'Chemistry.'" In Katherine Park and Lorraine Daston (eds), *The Cambridge History of Science, Vol. 3: Early Modern Science*. New York: Cambridge University Press.
Newman, William. 2014. "Robert Boyle, Transmutation, and the History of Chemistry Before Lavoisier: A Response to Kuhn." *Osiris*, 29: 63–77.
Newman, William. 2017. "Alchemical and Chymical Principles." In Peter Anstey (ed.), *The Idea of Principles in Early Modern Thought: Interdisciplinary Perspectives*. New York: Routledge.
Newman, William. 2018. *Newton the Alchemist: Science, Enigma, and the Quest for Nature's "Secret Fire."* Princeton, NJ: Princeton University Press.
Newman, William, and Lawrence Principe. 1998. "Alchemy vs Chemistry: The Etymological Origins of a Historiographic Mistake." *Early Science and Medicine*, 3: 32–65.
Newman, William, and Lawrence Principe. 2002. *Alchemy Tried in the Fire: Starkey, Boyle, and the Fate of Helmontian Chymistry*. Chicago, IL: University of Chicago Press.
Newman, William, and Lawrence Principe (eds). 2004. *George Starkey. Alchemical Laboratory Notebooks and Correspondence*. Chicago, IL: University of Chicago Press.
Newman, William, and Lawrence Principe. 2006. "Alchemy and the Changing Significance of Analysis." In Jed Z. Buchwald and Allan Franklin (eds), *Wrong for the Right Reasons*. Dordrecht: Springer.
Nummedal, Tara. 2002. "Practical Alchemy and Commercial Exchange in the Holy Roman Empire." In Pamela Smith and Paula Findlen (eds), *Merchants and Marvels: Commerce, Science, and Art in Early Modern Europe*. New York: Routledge.
Nummedal, Tara. 2007. *Alchemy and Authority in the Holy Roman Empire*. Chicago, IL: University of Chicago Press.
Nummedal, Tara. 2011. "Words and Works in the History of Alchemy." *Isis*, 102(2): 330–7.
Nummedal, Tara. 2013. "Introduction. Alchemy and Religion in Christian Europe." In Tara Nummedal, ed. Special Issue. *Alchemy and religion in Christian Europe*. *Ambix*, 60(4): pp. 311–22.
Nummedal, Tara. 2019. *Anna Zieglerin and the Lion's Blood: Alchemy and End Times in Reformation Germany*. Philadelphia: University of Pennsylvania Press.
Nummedal, Tara, and Donna Bilak (eds). 2020. *Furnace and Fugue: A Digital Edition of Michael Maier's Atalanta fugiens (1618) with Scholarly Commentary*. Charlottesville: University of Virginia Press.
Ogilvie, Brian W. 2006. *The Science of Describing: Natural History in Renaissance Europe*. Chicago, IL: University of Chicago Press.
Orenstein, Nadine M. 2001. *Pieter Bruegel the Elder: Drawings and Prints*. New Haven, CT: Yale University Press.
Osler, Margaret J. 2010. *Reconfiguring the World: Nature, God, and Human Understanding from the Middle Ages to Early Modern Europe*. Baltimore, MD: Johns Hopkins University Press.
Padilla, Carmela, and Barbara Anderson (eds). 2015. *A Red Like No Other: How Cochineal Conquered the World*. New York: Rizzoli.
Pagel, Walter. 1958. *Paracelsus: An Introduction to Philosophical Medicine in the Era of the Renaissance*. Basel: Karger.

Pagel, Walter. 1982a. *Paracelsus: An Introduction to Philosophical Medicine in the Era of the Renaissance*, 2nd ed. New York: Karger.

Pagel, Walter. 1982b. *Joan Baptista van Helmont: Reformer of Science and Medicine*. Cambridge: Cambridge University Press.

Pagel, Walter. 1986. *From Paracelsus to van Helmont: Studies in Renaissance Medicine and Science*. London: Variorum Reprints.

Pagel, Walter. 2002. *Joan Baptista van Helmont: Reformer of Science and Medicine*. Cambridge: Cambridge University Press.

Palmer, Richard. 1985. "Pharmacy in the Republic of Venice in the Sixteenth Century." In A. Wear, R.K. French, and I.M. Lonie (eds), *The Medical Renaissance of the Sixteenth Century*. Cambridge: Cambridge University Press.

Pantin, Isabelle. 2013. "John Hester's Translations of Leornardo Fioravanti: The Literary Career of a London Distiller." In S. Barker and B. Hosington (eds), *Renaissance Cultural Crossroads: Translation, Print and Culture in Britain 1473–1640*. Leiden: Brill.

Paracelsus. 1590a. *An excellent treatise teaching howe to cure the French-pockes*. London: Charlwood.

Paracelsus. 1590b. *De inventione artium*. In Johannes Huser (ed.), *Bücher und Schrifften*. Basel: Conrad Waldkirch.

Park, Katharine. 2011. "Observation in the Margins, 500–1500." In Lorraine Daston and Elizabeth Lunbeck (ed.), *Histories of Scientific Observation*. Chicago, IL: University of Chicago Press.

Parker, Goeffrey. 1988. *The Military Revolution: Military Innovation and the Rise of the West, 1500–1800*. Cambridge: Cambridge University Press.

Parkhurst, Charles. 1971. "A Color Theory from Prague, Anselm de Boodt, 1609." *Allen Memorial Art Museum Bulletin*, 29(1): 2–10.

Parshall, Karen Hunger, Michael T. Walton, and Bruce T. Moran (eds). 2015. *Bridging Traditions: Alchemy, Chemistry, and Paracelsian Practices in the Early Modern Era*. Kirksville, MO: Truman State University Press.

Partington, James R. [1960] 1999. *A History of Greek Fire and Gunpowder*. Baltimore, MD: Johns Hopkins University Press.

Partington, James R. 1961. *A History of Chemistry*. 4 vols. London: Macmillian.

Pelling, Margaret. 2003. *Medical Conflicts in Early Modern London: Patronage, Physicians, and Irregular Practitioners, 1550–1640*. Oxford: Oxford University Press.

Pereira, Michela. 2000. "Heavens on Earth. From the Tabula Smaragdina to the Alchemical Fifth Essence." *Early Science and Medicine*, 5: 131–44.

Pešek, Jiri. 1991. *The University of Prague, Czech Latin Schools, and Social Mobility (1570–1620)*. Oxford: Oxford University Press.

Plat, Hugh. 1593. *A briefe apologie of certaine nevv inventions*. London: Field.

Plat, Hugh. 1594. *The iewell house of art and nature Conteining diuers rare and profitable inuentions, together with sundry new experimentes in the art of husbandry, distillation, and moulding*. London: Peter Short.

Platt, Tristan. 2000. "The Alchemy of Modernity. Alonso Barba's Copper Cauldrons and the Independence of Bolivian Metallurgy (1790–1890)." *Journal of Latin American Studies*, 32: 1–53.

Pogliani, Annarita. 2009. "Il successo editoriale del *Büchlein von den ausgebrannten Wässern* di Michael Puff aus Schrick." In L. Vezzosi (ed.), *La letteratura tecnico-scientifica nel Medioevo germanico: Fachliteratur e Gebrauchstexte*. Alessandria: Edizioni dell'Orso.

Pomata, Gianna. 1998. *Contracting a Cure: Patients, Healers, and the Law in Early Modern Bologna*. Baltimore, MD: Johns Hopkins University Press.
Pomata, Gianna. 2011. "Observation Rising: Birth of an Epistemic Genre, 1500–1650." In Lorraine Daston and Elizabeth Lunbeck (eds), *Histories of Scientific Observation*. Chicago, IL: University of Chicago Press.
Pomeranz, Kenneth. 2000. *The Great Divergence: China, Europe, and the Making of the Modern World Economy*. Princeton, NJ: Princeton University Press.
Porta, Giambattista della. (1957, rept. London 1658). *Natural Magick*. Derek J. Price (ed.). New York: Basic Books.
Portuondo, Maria. 2009. *Secret Science: Spanish Cosmography and the New World*. Chicago, IL: University of Chicago Press.
Powers, John C. 2012. *Inventing Chemistry: Herman Boerhaave and the Reform of Chemical Arts*. Chicago, IL: Chicago University Press.
Priesner, Claus. 2011. "'Der zu vielen Wissenschaften anweisende curiöse Künstler': Alchemie, Volksmagie und Volksmedizin in barocken Hausbüchern." *Sudhoffs Archiv*, 95, 170–208.
Principe, Lawrence. 1990. "The Gold Process: Directions in the Study of Robert Boyle's Alchemy." In Z.R.W.M. Von Martels (ed.), *Alchemy Revisited: Proceedings of the International Conference on the History of Alchemy at the University of Groningen, 17–19 April 1989*. Leiden and New York: Brill.
Principe, Lawrence. 1998. *The Aspiring Adept: Robert Boyle and His Alchemical Quest*. Princeton, NJ: Princeton University Press.
Principe, Lawrence. 2000. "Apparatus and Reproducibility in Alchemy." In Frederic L. Holmes and Trevor Levere (eds), *Instruments and Experimentation in the History of Chemistry*. Cambridge, MA: MIT Press.
Principe, Lawrence. 2003. "Boyle's Alchemical Pursuits." In Michael Hunter (ed.), *Robert Boyle Reconsidered*. Cambridge: Cambridge University Press.
Principe, Lawrence (ed.). 2007. *Chymists and Chymistry: Studies in the History of Alchemy and Early Modern Chemistry*. Sagamore Beach, MA: Science History Publications.
Principe, Lawrence. 2007. "A Revolution Nobody Noticed? Changes in Early Eighteenth-Century Chymistry." In Lawrence M. Principe (ed.), *New Narratives in Eighteenth-Century Chemistry*. Dordrecht: Kluwer.
Principe, Lawrence. 2011. "Alchemy Restored," *Isis*, 102(2): 305–12.
Principe, Lawrence. 2013. *The Secrets of Alchemy*. Chicago, IL: University of Chicago Press.
Principe, Lawrence. 2014a. "The End of Alchemy? The Repudiation and Persistence of Chrysopoeia at the Académie Royale des Sciences in the Eighteenth Century." *Osiris*, 29: 96–116.
Principe, Lawrence. 2014b. "Goldsmiths and Chymists: The Activity of Artisans Within Alchemical Circles." In Sven Dupré (ed.), *Laboratories of Art: Alchemy and Art Technology from Antiquity to the 18th Century* (Archimedes, vol. 37). Cham: Springer.
Principe, Lawrence. 2014c. "A Practical Science: The History of Alchemy." In Sven Dupré, Dedo Von Kerssenbrock-Krosigk, and Beat Wismer (eds), *Art and Alchemy: The Mystery of Transformation*. Munich: Hirmer; Düsseldorf: Stiftung Museum Kunstpalast.
Principe, Lawrence M., and Lloyd DeWitt. 2002. *Transmutations. Alchemy in Art: Selected Works from the Eddleman and Fisher collections at the Chemical Heritage Foundation*. Philadelphia, PA: Chemical Heritage Foundation.
Principe, Lawrence, and William Newman. 2001. "Some Problems with the Historiography of Alchemy." In William Newman and Anthony Grafton (eds),

Secrets of Nature: Astrology and Alchemy in Early Modern Europe. Cambridge, MA: MIT Press.

Prinke, Rafal T. 1999. "The Twelfth Adept. Michael Sendivogius in Rudolfine Prague." In John Matthews, Ralph White, and Christopher Bamford (eds), *The Rosicrucian Enlightenment Revisited*. New York: Lindisfarne Books.

Prinke, Rafal T., and Mike A. Zuber. 2018. "Alchemical Patronage and the Making of an Adept: Letter of Michael Sendivogius to Emperor Rudolf II and His Chamberlain Hans Popp." *Ambix*, 65(4): 324–55.

Probert, Alan. 1969. "Bartolomé de Medina: The Patio Process and the Sixteenth Century Silver Crisis." *Journal of the West*, 8: 90–124.

Purš, Ivo. 2004. "Tadeáš Hájek of Hájek and His Alchemical Circle." In Ivo Purš and Vladimir Karpenko (eds), *Alchemy and Rudolf II: Exploring the Secrets of Nature in Central Europe in the 16th and 17th Centuries*. Prague: Artefactum.

Purš, Ivo, and Vladimír Karpenko (eds). 2016. *Alchemy and Rudolf II. Exploring the Secrets of Nature in Central Europe in the 16th and 17th Centuries*. Prague: Artefactum.

Ragland, Evan R. 2008. "Experimenting with Chymical Bodies: Reinier de Graaf's Investigations of the Pancreas." *Early Science and Medicine*, 13: 615–64.

Rampling, Jennifer. 2012. "John Dee and the Alchemists: Practising and Promoting English Alchemy in the Holy Roman Empire." *Studies in History and Philosophy of Science Part A*, 43: 498–508.

Rampling, Jennifer. 2014. "A Secret Language: The Ripley Scrolls." In Sven Dupré, Dedo von Kerssenbrock-Krosigk, and Beat Wismer (eds), *Art and Alchemy: The Mystery of Transformation*. Munich: Stiftung Museum Kunstpalast.

Rankin, Alisha. 2008. "Duchess, Heal Thyself: Elisabeth of Rochlitz and the Patient's Perspective in Early Modern Germany." *Bulletin of the History of Medicine*, 82(1): 109–44.

Rankin, Alisha. 2009. "Empirics, Physicians, and Wonder Drugs in Early Modern Germany: The Case of the Panacea Amwaldina." *Early Science and Medicine*, 14: 680–710.

Rankin, Alisha. 2011. "Germany." In J. Raymond (ed.), *The Oxford History of Popular Print Culture*. New York: Oxford University Press.

Rankin, Alisha. 2013. *Panaceia's Daughter: Noblewomen as Healers in Early Modern Germany*. Chicago, IL: University of Chicago Press.

Rankin, Alisha. 2014. "How to Cure the Golden Vein: Medical Remedies as Wissenschaft in Early Modern Germany." In Pamela Smith, Harold J. Cook, and Amy R.W. Meyers (eds), *Ways of Making and Knowing*. Ann Arbor, MI: University of Michigan Press.

Rattansi, P.M. 2004. "The Helmontian–Galenist Controversy in Restoration England." In Allen G. Debus (ed.), *Alchemy and Early Modern Chemistry Papers from Ambix*. Huddersfield: Jeremy Mills Publishing.

Ray, Meredith K. 2015. *Daughters of Alchemy: Women and Scientific Culture in Early Modern Italy*. Cambridge, MA: Harvard University Press.

Read, John. 1947. *The Alchemist in Life, Literature and Art*. London: Thomas Nelson.

Reinesius, Thomas. [1624] 1678. *Chimiatria, hoc est medicina, nobili et necessaria sui parte chimia, instructa et exornata* … Jenae: Excudit Johannes Gollnerus.

Rey Bueno, Mar. 2009. "La Mayson Pour Distiller Des Eaües at El Escorial: Alchemy and Medicine at the Court of Philip II, 1556–1598." *Medical History*, 53(S29): 26–39.

Rey Bueno, Mar. 2015. "If They Are Not Pages That Cure, They Are Pages That Teach How to Cure." In Karen Hunger Parshall, Michael T. Walton, and Bruce T. Moran (eds), *Bridging Traditions: Alchemy, Chemistry and Paracelsian Traditions in the Early Modern Era*. Kirksville, MO: Truman State University Press.

Roberts, Lissa. 2005. "The Death of the Sensuous Chemist: The 'New' Chemistry and the Transformation of Sensuous Technology." In David Howes (ed.), *Empire of the Senses: The Sensual Culture Reader*. Oxford: Berg.

Roberts, L., S. Schaffer, and P. Dear (eds). 2007. *The Mindful Hand: Inquiry and Invention from the Late Renaissance to Early Industrialisation*. Amsterdam: Royal Netherlands Academy of Arts and Sciences.

Rocke, Alan J. 1985. "Agricola, Paracelsus, and 'Chymia.'" *Ambix*, 32: 37–45.

Roos, Anna Marie. 2007. *The Salt of the Earth: Natural Philosophy, Medicine, and Chymistry in England, 1650–1750*. Leiden: Brill.

Rorty, Amélie Oksenberg. 1988. "Persons and Personae." In Amélie Oksenberg Rorty (ed.), *Mind in Action: Essays in the Philosophy of Mind*. Boston, MA: Beacon Press.

Rosetti, Giovanventura. [1555] 1973. *Notandissimi secreti de l'arte profumatoria*. Rome: Neri Pozza.

Röslin, Eucharius. 1533. *Kreutterbuoch von allem Erdtgewaechs*. Frankfurt: Egenolff.

Rublack, Ulinka. 2002. "Fluxes: The Early Modern Body and the Emotions." *History Workshop Journal*, 53(1): 1–16.

Ryff, Walther Hermann. 1545. *Das New groß Distillier Buoch Wolgegründter Künstlicher Distillation …* Frankfurt: Christian Egenolff.

Sadler, Lynn Veach. 1975. "Alchemy and Greene's Friar Bacon and Friar Bungay." *Ambix*, 22(2): 111–24.

Sala, Angelus. 1617. *Anatomia vitrioli*. Leiden: Basson.

Sala, Angelus. 1622. *Chrysologia, seu examen auri chymicum*. Hamburg: Carstens.

Schama, Simon. 1987. *The Embarrassment of Riches*. New York: Knopf.

Schenda, R. 1982. "Der 'gemeine Mann' und sein medikales Verhalten im 16. und 17. Jahrhundert." In Joachim Telle (ed.), *Pharmazie und der gemeine Mann: Hausarznei und Apotheke der frühen Neuzeit*. Braunschweig: Waisenhaus Verlag.

Schmitt, Charles B. 1969. "Experience and Experiment: A Comparison of Zabarella's View with Galileo's in De Motu." *Studies in the Renaissance*, 16: 80–138.

Schrick, Michael. 1476. *Von den ausgebrannten Wassern*. Augsburg: Bämler.

Schroder, Johann. 1669. *The Compleat Chymical Dispensatory … Written in Latin by Dr. John Schroder … and Englished by William Rowland …* London: John Darby for Richard Clavell.

Sennert, Daniel. 1619. *De chymicorum cum Aristotelicis et Galenicis consensu ac dissensu*. Wittenberg: Zacharias Schürer.

Sennert, Daniel. 1676. "Epistolarum Medicinalium Centuriae Duae." In *Danielis Sennerti Vratislavensis … operum in sex tomos divisorum*, vol. 6. Lyon: Jean-Antoine Huguetan.

Sennett, Richard. 2008. *The Craftsman*. New Haven, CT: Yale University Press.

Shackelford, Jole. 1993. "Tycho Brahe, Laboratory Design, and the Aim of Science: Reading Plans in Context." *Isis*, 84(2): 211–30.

Shackelford, Jole. 1998. "Seeds with a Mechanical Purpose: Severinus' Semina and Seventeenth-Century Matter Theory." In Allen G. Debus and Michael T. Walton (eds), *Reading the Book of Nature: The Other Side of the Scientific Revolution*. Kirksville, MO: Sixteenth Century Journal Publishers.

Shackelford, Jole. 2004. *A Philosophical Path for Paracelsian Medicine: The Ideas, Intellectual Context, and Influence of Petrus Severinus (1540–1602)*. Copenhagen: Museum Tusculanum Press.
Shadwell, Thomas. 1676. *The Virtuoso: A Comedy, acted at the Duke's Theatre*. London: Printed by T.N. for Henry Herringman.
Shapin, Steven. 1984. "Pump and Circumstance: Robert Boyle's Literary Technology." *Social Studies of Science*, 14: 481–520.
Shapin, Steven, and Simon Schaffer. 1985. *Leviathan and the Air-Pump: Hobbes, Boyle, and the Experimental Life*. Princeton, NJ: Princeton University Press.
Smith, Cyril Stanley. 1988. *A History of Metallography: The Development of Ideas on the Structure of Metals before 1890*. Cambridge, MA: MIT Press.
Smith, Pamela. 1994. *The Business of Alchemy: Science and Culture in the Holy Roman Empire*. Princeton, NJ: Princeton University Press.
Smith, Pamela. 2004. *The Body of the Artisan: Art and Experience in the Scientific Revolution*. Chicago, IL: University of Chicago Press.
Smith, Pamela. 2006. "Laboratories." In Lorraine Daston and Katharine Park (eds), *The Cambridge History of Science, Vol. 3: Early Modern Europe*. Cambridge: Cambridge University Press.
Smith, Pamela. 2011. "What Is a Secret? Secrets and Craft Knowledge in Early Modern Europe." In Elaine Leong and Alisha Rankin (eds), *Secrets and Knowledge in Medicine and Science, 1500–1800*. Farnham: Ashgate Publishing, Ltd.
Smith, Pamela. 2014. "Knowledge in Motion: Following Itineraries of Matter in the Early Modern World." In Daniel Rogers, Bhavani Raman, and Helmut Reimitz (eds), *Cultures in Motion*. Princeton, NJ: Princeton University Press.
Smith, Pamela. 2016. "Historians in the Laboratory: Reconstruction of Renaissance Art and Technology in the Making and Knowing Project." *Art History*, 39: 210–33.
Soukup, R. Werner. 2007. "Crucibles, Cupels, Curcurbits: Recent Results of Research on Paracelsian Alchemy in Austria around 1600." In Lawrence M. Principe (ed.), *Chymists and Chymistry: Studies in the History of Alchemy and Early Modern Chemistry*. Sagamore Beach, MA: Watson Publishing.
Spargo, Peter. 2005. "Investigating the Site of Newton's Laboratory in Trinity College, Cambridge." *South African Journal of Science*, 101: 315–21.
Sparling, Andrew W. 2018. "Providence and Alchemy: Paracelsus on How Knowledge Unfolded, Matter Developed, and Bodies Might Be Perfected." Ph.D. thesis, University of Nevada, Reno.
Stahl, Georg Ernst. 1697. *Zymotechnia fundamentlis, seu, Fementationis theoria generalis, qua nobilissimæ hujus artis & partis chymiæ, utilissimæ atq[ue] subtilissimæ, causæ & effectus in genere, ex ipsis mechanico-physicis principiis, summo studio eruuntur, simulque experimentum novum sulphur verum arte producendi, et alia utilia experimenta atque observata, inseruntur*. Halle an der Saale: Christoph Salfeld.
Stahl, Georg Ernst. 1723. *Fundamenta Chymiae Dogmaticae & Experimentalis*. Nuremberg: Mauritius.
Starkey, George. 1658. *Pyrotechny Asserted and Illustrated, To be the surest and safest means for Arts Triumph over Natures infirmities*. London: For Samuel Thomson.
Stein, Claudia. 2009. *Negotiating the French Pox in Early Modern Germany*. Farnham: Ashgate Publishing, Ltd.
Storey, Tessa. 2011. "Face Waters, Oils, Love Magic and Poison: Making and Selling Secrets in Early Modern Rome." In Elaine Leong and Alisha Rankin (eds), *Secrets and Knowledge in Medicine and Science, 1500–1800*. Farnham: Ashgate Publishing.

Stuart, Kathy. 1994. "The Executioner's Healing Touch." In Max Reinhart (ed.), *Infinite Boundaries: Order, Disorder, and Reorder in Early Modern German Culture*. Kirksville, MO: Sixteenth Century Journal Publishers.

Sumption, Jonathan. 2017. *The Hundred Years War. Vol. 4: Cursed Kings* (The Middle Ages Series). Philadelphia: University of Pennsylvania Press.

Szőnyi, György. 1997. "Scientific and Magical Humanism at the Court of Rudolf II." In Eliška Fucíková (ed.), *Rudolf II and Prague: The Court and the City*. Prague: Prague Castle Administration; London: Thames and Hudson.

Taape, Tillmann. 2014. "Distilling Reliable Remedies: Hieronymus Brunschwig's *Liber de arte distillandi* (1500) between Alchemical Learning and Craft Practice." *Ambix*, 61: 236–56.

Telle, Joachim (ed.). 1994. *Analecta Paracelsica: Studien zum Nachleben Theophrast von Hohenheims im deutschen Kulturgebiet der frühen Neuzeit*. Stuttgart: F. Steiner.

Tilton, Hereward. 2003. *The Quest for the Phoenix: Spiritual Alchemy and Rosicrucianism in the Work of Count Michael Maier (1569–1622)*. Berlin: Walter de Gruyter.

Tilton, Hereward. 2020. "Michael Maier: An Itinerant Alchemist in Late Renaissance Germany." In Tara Nummedal and Donna Bilak (eds), *Furnace and Fugue: A Digital Edition of Michael Maier's Atalanta fugiens (1618) with Scholarly Commentary*. Charlottesville: University of Virginia Press.

Timmermann, Anke. 2008. "Doctor's Order: An Early Modern Doctor's Alchemical Notebooks." *Early Science and Medicine*, 13(1): 25–52.

Van Helmont, Jan Baptiste. 1644. *Opuscula medica inaudita*. Cologne: Apud Jodocum Kalcoven.

Van Helmont, Jan Baptiste. 1652. *Ortus medicinae*. Amsterdam: Elzevir.

Van Helmont, Jan Baptiste. 1664. *Van Helmont's works containing his most excellent philosophy, physick, chirurgery, anatomy …* London: Lloyd.

Van Helmont, Jan Baptiste. 1667. *Ortus medicinae*. Lyon: Huguetan and Barbier.

Van Helmont, Jan Baptiste. 1966. *Ortus medicinae*. Amsterdam: Elsevir. Reprint Brussels: Culture et Civilisation.

Van Thiel-Stroman, Irene, and Pieter van Thiel. 2006. *Painting in Haarlem 1500–1850: The Collection of the Frans Hals Museum*. Ghent: Ludion Ghent.

Vaughan, Thomas. 1650. *Anima magica abscondita, or, A Discourse of the Universall Spirit of Nature, with His Trange, Abstruse, Miraculous Ascent and Descent*. London: H. Blunden.

Venel, Gabriel Francois. 1753. "Chymie, ou Chimie." In Denis Diderot and Jean-Baptiste le Rond d'Alembert (eds), *Encyclopédie ou dictionnaire raisonné des sciences, des arts et des métiers*, vol. 3. Paris: s.n.

Von Hohberg, Wolf Helmhardt. 1687. *Georgica curiosa aucta, das ist, Umständlicher Bericht und klarer Unterricht von dem adelichen Land- und Feld-Leben*. Nuremberg: The Heirs of Johann Friedrich Endter.

Walton, Michael. 2011. *Genesis and the Chemical Philosophy: True Christian Science in the Sixteenth and Seventeenth Centuries*. New York: AMS Press.

Watanabe O'Kelly, Helen. 2002. "Saxony, Alchemy and Dr. Faustus." In Alexandra Lembery and Elmar Schenkel (ed.), *The Golden Egg: Alchemy in Art and Literature*. Berlin: Glienicke.

Webster, Charles. 1977. *The Great Instauration: Science, Medicine and Reform 1626–1660*. London: Duckworth.

Webster, Charles. 2008. *Paracelsus: Medicine, Magic and Mission at the End of Time*. New Haven, CT: Yale University Press.

Webster, John. 1654. *Academiarum examen, or the examinatio of academies*. London: Giles Calvert.

Weeks, Andrew. 1997. *Paracelsus: Speculative Theory and the Crisis of the Early Reformation*. Albany: State University of New York Press.

Weeks, Andrew. 2008. *Paracelsus (Theophrastus Bombastus von Hohenheim, 1493–1541): Essential Theoretical Writings*. Boston, MA: Brill.

Westermann, Mariet. 1997. *The Amusements of Jan Steen: Comic Painting in the Seventeenth Century*. Zwolle: Waanders Publishers.

Westman, Robert S. 2011. *The Copernican Question: Prognostication, Skepticism, and Celestial Order*. Berkeley and Los Angeles: University of California Press.

Wheelock, Arthur. n.d. Cornelis Bega/The Alchemist/1663. In *Dutch Paintings of the Seventeenth Century*, NGA Online Editions. Available online: http://purl.org/nga/collection/artobject/161648.

Williams, Alan. 1975. "The Production of Saltpetre in the Middle Ages." *Ambix*, 22: 126–33.

Willis, Thomas. 1659. *Diatribae duae medico-philosophicae quarum prior agit de fermentation …* London: Roycroft.

Willis, Thomas. 1660. *Diatribae duae medico-philosophicae quarum prior agit de fermentation …* London: Roycroft.

Woodward, Walter W. 2010. *Prospero's America: John Winthrop, Jr., Alchemy, and the Creation of New England Culture, 1606–1676*. Chapel Hill: University of North Carolina Press.

Zetzner, Lazarus, and His Heirs (eds). 1602–1661. *Theatrum chemicum*. 6 vols. Strasbourg: Lazarus Zetzner and His Heirs.

Zilsel, Edgar. 1942. "The Sociological Roots of Science." *American Journal of Sociology*, 47: 544–62.

Zilsel, Edgar. 2003. *The Social Origins of Modern Science*. Diederick Raven, Wolfgang Krohn, and Robert S. Cohen (eds). Dordrecht and Boston, MA: Kluwer.

Ziolkowski, Theodore. 2015. *The Alchemist in Literature: From Dante to the Present*. Oxford: Oxford University Press.

LIST OF CONTRIBUTORS

Donna Bilak is an independent scholar and historian of early modern alchemy, specializing in the study of emblematics and laboratory processes to examine how text, images, and the experimental use of materials come together in the creation and application of knowledge in the early modern period. Her research interests also extend to jewelry history and technology, which draws upon her previous professional experience in Toronto's jewelry industry as a designer and wax model-maker. She received her Ph.D. from the Bard Graduate Center and is the Director of 12 Keys Consultancy & Design, LLC.

Elisabeth Berry Drago studies interconnected histories of art and science in the Dutch Golden Age, as well as the representation of women in artisanal practices. She received her PhD from the University of Delaware and is a Research Curator at the Science History Institute in Philadelphia. Berry Drago is the author of *Painted Alchemists: Early Modern Artistry and Experiment in the Work of Thomas Wijck* (2019).

William Eamon is Regents Professor of History Emeritus at New Mexico State University. A historian of science, he has published on a variety of topics concerning medieval and early modern science, medicine, and technology. He is the author of *Science and the Secrets of Nature: Books of Secrets in Medieval and Early Modern Culture* (1994) and *The Professor of Secrets: Mystery, Magic, and Alchemy in Renaissance Italy* (2010). He has also published on the history of magic, astrology, and the occult sciences in early modern Europe. He is currently writing a book, *Science and Everyday Life in Early Modern Europe*.

Margaret D. Garber is an Associate Professor of History of Science at California State University Fullerton. She has authored a number of articles and edited volume chapters on the history of alchemy, iatrochemistry, and scientific societies in the Holy Roman Empire. Currently, she is completing a monograph on the *Academia Naturae Curiosorum* (aka *Leopoldina*) entitled "Domesticating Curiosities: The *Leopoldina*'s Literary Transformation of Chemistry and Medicine in the Holy Roman Empire (1650–1750)."

Joel A. Klein stewards the history of medicine and pre-1800 science collections at The Huntington Library. He completed his Ph.D. in the History and Philosophy of Science at Indiana University and has had postdoctoral positions at Columbia University and the Science History Institute. Dr. Klein's study of early modern medicine and science centers on the intersections between medicine, chemistry, and matter theory in seventeenth-century Germany. He is writing a book on *Chymical Life in Early-Modern Europe*, was recently appointed book reviews editor for *Early Science and Medicine*, and works on several digital projects, including "The Chymistry of Isaac Newton."

Bruce T. Moran is Professor Emeritus at the University of Nevada, Reno. Along with many other books and articles he is the author of *Distilling Knowledge* (2005), *Andreas Libavius and the Transformation of Alchemy* (2007), and, most recently, *Paracelsus: An Alchemical Life* (2019). He is also coeditor of *Bridging Traditions: Alchemy, Chemistry, and Paracelsian Practices in the Early Modern Era* (2015). His current project focuses on "The Alchemy of the Potters' Art in Early Modern Italy."

Lawrence M. Principe is the Drew Professor of Humanities at Johns Hopkins University in the Departments of History of Science and Technology and of Chemistry. His research focuses on late medieval and early modern alchemy/chemistry and the interactions of science and religion. His recent books include *The Scientific Revolution: A Very Short Introduction* (Oxford, 2011), *The Secrets of Alchemy* (Chicago, 2013), and *The Transmutations of Chymistry: Wilhelm Homberg and the Académie Royale des Sciences* (Chicago, 2020). He has received the Francis Bacon Medal (2005) and the Prix Franklin-Lavoisier (2016) for his scholarly contributions.

Andrew Sparling is an independent scholar living in Providence, Rhode Island. He holds a Ph.D. from the University of Nevada, Reno. His research explores the roles that religion, natural-philosophical theory, and laboratory investigation played in science from the sixteenth to the eighteenth centuries. The chief focus of his recent work has been the alchemy of Paracelsus.

Tillmann Taape is a historian of early modern science and medicine. His research is on alchemy, craft knowledge, and the history of early print, with a particular interest in the relationship between learned and vernacular traditions of knowledge. In addition to traditional scholarship, he pursues the reconstruction of historical techniques and processes as a research tool in his work with the Making and Knowing Project at Columbia University.

INDEX

Aachen, Hans von 179
academies 15–16
Agrippa, Cornelius 1
Agricola, Georg 46, 99, 106, 123, 137–8, 142, 158–9, 183
Alba, Duke of 145
"alchemist", use of the term 110, 128
alchemists
    attitudes to 200–2, 209–13
    imagery of 208
    status of 203, 211
    suspicion of 203, 206
    in text 202–8
alchemy 6–11, 16, 20, 98, 106
    caricatured view of 42
    change in nature of 43
    golden age of 200
    identity of 44
    power to transform and to generate new materials 199
    traditional form of 93, 99–104, 155
    treated as a commodity 142
    uses of term 24–5, 36
    workaday 142
    writ large 100
    see also representations
alcoholic spirits 115
allegorical schemes 86
Allin, John 71, 82–4, 88
Amsterdam 6
analysis and analytical reasoning 18–19, 65
Ancel, Guillaume 178, 183
Andreae, Johann Valentin 206
Andrew, Laurence 112
Anna of Saxony, Electress 149
anthologies 105
Antwerp 6
apothecaries 115–17, 123
apparatus, chemical 74, 150–1, 202, 209–10
aquavitae brothers 114–15
Aquinas, Thomas 43, 189
Aristotle and Aristotelianism 11, 20, 26–7, 33–4, 37, 48, 53–6, 99, 159–60, 173, 175, 189
Arndt, Johann 104
Arriaga, Roderigo 193
artisans
    expertise of 92–3, 120, 155, 169
    social distance from scholars 94
    "superior" 140
Ashmole, Elias 194, 205
astronomical observation 174
Atalanta fugiens 84–7
August, Elector of Saxony 15
August the Younger of Braunschweig and Lüneberg, Duke 4–5
Avery, William 70
Avicenna 10, 48

Bachachius, Martin 192
Bacon, Francis 8, 17–19, 91, 143, 168
Bacon, Roger 43, 205–6

Baker, George 166
Banister, John 166
Barba, Álvaro Alonso 136
Barnaud, Nicolas 178
Basan, Pierre François 221
Basel University 175
"Basil Valentine" 87, 177–80, 185
Becher, Daniel 197
Becher, Johann Joachim 6–7, 29–30, 143, 184–5
Bega, Cornelis 213, 220–4
Beguin, Jean 56, 194
Benedict, Barbara 203
Bernardoni, Andrea 142
Bertereau, Martine de 107
Biggs, Noah 62
Billich, Anton Günther 56
Biringuccio, Vannoccio 116–17, 123, 128, 142, 158–9, 162
Black Death 115
Bodenstein, Adam von 12
body, human, as a microcosm 172, 176
Boehme, Jakob 104
Boerhaave, Herman 196–8
Bohn, Johannes 196
Bonsisti, Sisto de 127
"book of nature" 83
books 4, 105–7, 124–5, 146, 150, 155–6, 161, 165–9
Borgarucci, Prospero 123
Bostocke, Richard 51–3, 62
Botero, Giovanni 143
Boyle, Robert 8, 14–17, 20, 30–8, 55, 57, 64, 69–70, 84, 99, 102, 109, 131, 134–5, 201, 206
Bragadino, Marco 130
Brahe, Tycho 69, 179, 183–4, 188
Brakenburgh, Richard 213–16, 226–7
brandy 145–6, 152
Brant, Sebastian 128, 202, 205
Brendel, Zacharius Junior 14
Brendel, Zacharius Senior 190–1
Bruegel, Pieter the Elder 73–5, 88, 208–9, 213–16, 226–7
Brune, Johan de 216
Bruno, Giordano 200
Brunschwig, Hieronymus 48–9, 112, 123, 145–61, 168
Bulkeley, Gershom 71, 80–4, 88
Bullinger, Heinrich 97

cabinets of curiosity 15
calcination 69
Cambridge University 196
Campanella, Tommaso 16
capitalism 6
Cardanus, Hieronymus 10
Cavendish, Margaret 227
cementation 44–7
Cennini, Cennino d'Andrea 119
Cesi, Federico 15
change, unprecedented nature of 2, 6
charcoal 138
charlatan image 101
Charles II of England 200
Charles II of Spain 180, 187
Chaucer, Geoffrey 202, 205
Chauliac, Guy de 224
chemia 11–12
chemical arts 13–14, 142–4, 149, 155–6, 160–3, 166–9
  popularization of 125
chemical industries 111, 130
  health hazards of 137
  worldwide impact of 139
chemical substances, qualities and properties of 24
chemistry
  definition of 1–2, 9–10, 21
  emergence as an autonomous discipline 11, 24, 65, 73, 110, 113, 142
  expansion and diversification of 23
  independence from medicine 11
  modern science of 91–2
  participation in culture of 107–8
  in pharmacies and marketplaces 115–16
  use of the word 142
chemists, identity and image of 1, 128–30
Christendom 90
Christian I of Anhalt-Bernburg 12
Christianity 4, 8, 48
chymiatria, definition of 11
chymical communities 173
chymistry 9–10, 25–38, 44, 47, 51–7, 60–5, 67–71, 84, 87, 120–6, 134–6, 139–40, 171–2, 177–80, 184, 193–5
  definition of 56, 63
  everyday form of 120–1, 140, 152
  globalization of 134–6
  golden age of 172–3, 177

practice of 108–11
in print 121–4
speculative 103–4
suspicion of 68
theories and concepts in 38
transmutational 35–8
at universities in Europe 186–98
Clave, Etienne de 28
climate change 138
Clowes, William 166
cochineal 118–19, 134–5
Cock, Hieronymus 208
coded language, use of 44, 84
Collaert, Jan 127, 211
color-men 117–19
Columbus, Christopher 4, 90–1
commercialization 155
composition, theories of 26–35
compound substances 26
Conring, Hermann 9
Cooper, William 155
Copernicus, Nicolaus 4–5, 90–1
copper 35
Corbett, Jane Russell 224
corpuscularian theories 20, 20, 32–4, 38, 61
Cortese, Lady Isabela 110, 119
Counter-Reformation 172
courts, noble 126–7, 171–3, 176, 198
craft knowledge 120–3, 125, 161
Croll, Oswald 12, 175, 180–3
Culpeper, Cheney 16
cupellation 44, 46
curiosity 14–16, 206–7
Cyriacus, Jacobus 105

"dearth science" 167–8
Debus, Allen 109
decorative arts 43, 179
deductive method 18
Dee, John 182, 200, 205
deforestation 138
del Rio, Martin 101–2
Democritus 59
Denmark 138
Descartes, René 8, 18, 32
Dia aromatica 164
Dickinson, Edmund 201
di Farre de Brescia, Giovanni Andrei 114

diseases, occupational 137–8
distillation 10–11, 47–8, 53, 68, 111–16, 119, 123, 130, 139, 142–64, 167–9
in the German lands 145–9
golden age of 111–14
of wine 145
distilled waters 146, 153–5, 158
Doria, Gian Andrea 130
Döring, Michael 57–60
Dorn, Gerhard 100, 103–5, 175
Du Chesne, Joseph 12, 28, 52–6, 62, 175–8
Dupré, Sven 161
Dürer, Albrecht 202
dyeing 117–18, 134
Dymock, Cressy 91

Eamon, Wlliam 48
East India Company, English 8
East India Company, Dutch 6
Egenolff, Christian 100–1, 122, 156
Eglinus, Raphael 186–7
Eisenstein, Elizabeth 104
electrum 10
elements (of fire, air, water and earth) 26, 47, 52
elixirs 10
Elizabeth I, Queen 183
emblems 84–8
Enlightenment thinking 42
Epicurus 32
epidemics 114
equipment, chemical *see* apparatus
Erasmus 97, 128
Ercker, Lazarus 123, 132–3, 142, 199
El Escorial 126
etchings 117–18
Europeanness 89
Evelyn, John 14, 138
experimentation 17–18, 43–4, 68, 173–4, 198

Fabri de Peiresc, Nicolas-Claude 16
Faust, story of 207–8
Ferdinand II, Elector 192
Ferdinand III, Emperor 178
Fioravanti, Leonardo 125, 127, 137, 145, 164–6
four-element theory 26–7
Foxcraft, Ezekiel 16

France 138
Francesco I, Grand Duke of Tuscany 211
François de St-André 33
fraud 128–30, 202–3, 208, 213
Frederick V, Elector 177, 192
Freke, Elizabeth 161
French, John 16, 62–3
Friedrich I of Württemberg, Duke 183–4
fumes, toxic 137
furnaces 75–80

Galen and Galenic medicine 12, 48, 53–4, 68, 97–8, 159–60, 163, 173, 176
Galileo 16, 90–1
Garzoni, Tomaso 116, 125, 130
Gassendi, Pierre 16, 32
Geber and Pseudo-Geber 10, 27, 44–7, 54–9, 110, 209–10
genera 9–10
Genoa 134
genre paintings 139, 200
Germany and German language 97, 105–6, 116, 122, 145–50, 155–60, 172–5, 183, 207
Gessner, Conrad 156, 158, 166
Glauber, Johann Rudolph 55, 100, 104, 195
Glisson, Francis 63–5
Glorez, Andreas 107–8
gold 35
    philosophical 100
    potable 60, 115
gold-making 201–2
goldsmithing 116
Graaf, Reinier de 196
Granvill, Joseph 201
Gratarolo, Guglielmo 105
Greene, Robert 205
Grüninger, Johann 156
guaiacum 163
Gualdi, Pietro 136
guilds 203
gunpowder 132–3, 158

Hájek, Tádeas 182–3
Harkness, Deborah 168
Hartlib, Samuel 16, 18, 91, 93, 168, 201
Hartmann, George 161
Hartmann, Johannes 57, 187, 189

Harvey, Gabriel 166
Heerschop, Hendrick 129, 213, 216–17
Henry IV, King of France 177–8
Hermannus, Philippus 165
Hermes Trismegistus 9–10
Hessen-Kassel 126, 176–8, 180
Hester, John 145, 164–9
Hill, Thomas 165–6
Hippocrates 12, 154 (and Hippocratic writing) 52
historians of science 92, 173
historiography of early modern science 42–3
Hoefnagel, Jacob 179
Hoffmann, Friedrich 196–7
Hofmann, Caspar 12
Hohenheim, Theophrastus von *see* Paracelsus
Hohberg, Wolf Helmhardt von 101
Honhauer, Georg 127
Honrick, Gerard 120–1, 132
Hooke, Robert 91, 206
Hooykaas, Reijer 56
Horcicky, Jacob 180
Houbraken, Arnold 224
household literature 107
household tasks 227
humors, bodily 97–8, 153, 157, 176
hypostasis 54
hylomorphism 26

iatrochemistry 11, 171–2, 175, 186, 195–8
inflation 6, 139
instruments, chemical *see* apparatus
interaction
    between those with chemical knowledge 93–4
    between those with direct sensory experience of nature 173
iron and ironsmithing 35, 117

Jacob, Margaret 98–9
Jamnitzer, Christof 179
Jardin Royal des Plantes, Paris 93
Jena, University of 190–1
Jesenius, Jan 188
Jesuati 114
Johann-Friedrich of Sachsen-Gotha, Duke 142

John XXI, Pope *see* Petrus Hispanus
Jonson, Ben 203–4
Juana of Habsburg, Infanta 149
Julius of Braunschweig-Wolfenbüttel, Duke 142, 183–4
Jungius, Joachim 57
Jüngken, Johann Heinrich 160

Kaufmann, Thomas DaCosta 208
Kelley, Edward 182, 200
Kepler, Johannes 179
Khunrath, Heinrich 21, 104, 200, 205–6, 227–8
Klein, Ursula 188
knowledge, making of 41, 92, 140
Krafftheim, Crato von 12
Kranich, Burchard 167
Kunckel, Johann 94–5
Kunstbüchlein 122–3, 160

labelling 154
laboratories 67–88, 125–8, 139, 180, 183–5, 188
  allegorical 83–7
  in paintings 213–27
laboratory processes 71
language, obscurity of 91
Latin language 93, 97, 150, 152, 156
Lavoisier, Antoine Laurent 39, 61, 92
Le Mort, Jacob 195–6
lead 35
learning about chemistry 124
Lee, Francis 133
LeFèvre, Nicaise 63, 75–83, 86, 88, 200–1
Leibärzte 174, 177, 182, 198
Leibniz, Gottfried Wilhelm 57
Leiden University 194–8
Lemery, Nicolas 33
Leonardo da Vinci 2, 142
Leopold I, Emperor 178
Levens, Peter 161
Libavius, Andreas 10–11, 54–6, 59, 69, 175, 190–1, 209–10
libraries 4
local setting for everyday life 89
London 166, 169
Lonicer, Adam 156, 158
Lucretius 56
Lull, Ramon 110

Luther, Martin 4, 106, 188, 207–8
Luther, Paul 188
Luyken, Jan and Caspar 211–12

Machiavelli, Nicolò 130
"Maddalena" 120
Maets, Carel de 195–6
magic 1–2, 107–8
Maier, Michael 83–8, 104, 142–3, 182, 200
Mansfeld, Dorothea von 146, 149
manufacturing 6–7, 111, 161
Marburg University 186–8
Marci, Jan Marcus 193–3
Margraf, Christoph 179
market forces 144
marketplaces 144
Markham, Gervase 119
Marx, Karl 7
material culture 43
matter, theories of 26, 42, 47, 54–5, 64–5, 70–1, 93, 99, 109, 143
Mattioli, Pier Andrea 11
Mayerne, Turquet de 178
Medici, Antonio 177
Medici, Ferdinand I de 15
Medici, Francesco I de 15, 126–7, 177
Medici, Leopoldo de' 16
medicine 11–14, 24, 69, 144–6, 149, 153–5, 158–69, 172–6
  allied to alchemy 47
  transformed by alchemy 42
Medina, Bartolomé de 117, 136
mercury 26–9, 48, 52–4, 61, 68–9, 84, 98, 175–6
Mersenne, Marin 16
metallurgy 7, 116–17, 138, 183
metalworking 158
Michelangelo 2–3
migrant labor 133
minima naturalia 27
mining 137, 143, 183
Miseroni, Ottavio 179
Mitchell, Jonathan 71
modernity 139
money 7
monism 30–2
Moran, Bruce (editor) 110, 176–7, 187
Morgenstern, Philipp 105

Moritz of Hessen-Kassel, Prince 126, 176–8, 180, 185–8
Mukherjee, Ayesha 167
Müller, Johann 182
multiple versions of a text 104–5
Mylius, Johann Daniel 197–8

national identity 89
natural knowledge 1, 7–8, 25, 52, 93, 156, 200, 206
natural philosophy 18–21, 43, 57, 90, 94, 99–101, 159–60, 172–3
natural world, the
    experienced through the senses 173
    "negative-empirical" principle applied to 53
    perceptions of 139
    study of 174
    understanding of and interaction with 41
Neoplatonic–Hermetic philosophies 174
Netherlands, the 6
Neuburg, Duke of 12
"new historiography of alchemy" 43
"new science" 41–2
Newman, William 47, 54, 57–62
Newton, Sir Isaac 8, 38, 70, 87
"noble" and "ignoble" metals 35
Nummedal, Tara 142, 184

occult sciences 179
Oldenburg, Henry 17
Ottoman Empire 90
Oxford University 194

Pacioli, Lucca 2
painters 117–19
paintings of laboratories 213–27
Palgrave, Richard 70
Palissy, Bernard 168
Paracelsus, portrait of 96
Paracelsus (and Paracelsian practice) 2, 4, 10–13, 20, 25–9, 48, 51–6, 59–62, 68–9, 87, 95–100, 103, 109, 115, 133, 137–8, 159–60, 165, 168, 175–8
Paris, University of 193–4
Park, Katharine 173–4
patio process 136

patronage 126–7
Paul of Taranto 44, 46
"perfect" substances 100
Perna, Peter 105
perspective drawing 2–4
Petrarch 128
Petrus Hispanus 161
Peucer, Caspar 188
pharmacists 115, 123
Philip II, King of Spain 125–7, 135, 180, 187
philosophers' stone 36–7, 42, 68, 83, 84, 102, 109–10, 177, 182, 185, 201
philosophy, traditional 52
phlogiston theory 7, 30
Piemontese, Alessio 125
Plat, Hugh 145, 167–9
Plattes, Gabriel 16
"pleasing novelty" of chemical practice 162
"plusquamperfect" elixir 38
pollution 137
Popp, Johann 16
Porta, Giambattista della 2–3, 16
the pox 162–3
practical aspect of chemistry 42–3
Prague 180, 183
    Carolinum University 191–3
prime matter 26–7
prince-practitioners 177–8, 185, 193
Principe, Lawrence 47, 61–2, 86, 111, 185, 216
printing 4, 48, 104–7, 112, 116–17, 122, 125, 145, 155, 160–1, 169, 210–11
probatum est phrase 124
projection process 37
Protestant Reformation 4, 90, 172
purification of metals 46–8

Quercetanus *see* Du Chesne, Joseph
"quintessences" 103, 153

Ranelagh, Lady 201
Rankin, Alisha 146
Read, John 211
recipes 124–5, 161–2
"reduction to the pristine state" 34, 47
religious orders 114
Renaissance thinking 2, 173–4

representations of alchemy and alchemy's ideas 227
Richelieu, Cardinal 107
Ripley, George 205
Ripley Scrolls 86
ritual 121
Rodovský, Bavor 200
Rolfinck, Werner 191
Rome, city of 6
Rosetti, Giovanni Ventura 112, 124
Röslin, Eucharius 156
Röslin, Helisaeus 54
Rossi, Girolamo *see* Rubeus, Hieronymus
royal residences as seedbeds for chymical knowledge 173
Royal Society of London 14–18, 41, 168, 200–3, 206
Rubeus, Hieronymus 10
Rudolf II, King of Bohemia and Holy Roman Emperor 126, 178–3, 200
Rupescissa, John of 47–8, 60, 103, 114, 152, 158, 160
Ruscelli, Girolamo 161
Ryff, Walther Hermann 156–7, 163, 165

Sadeler, Aegidius 180
Sala, Angelus 55–6, 61
salt 27, 29, 48, 52–3, 61, 68–9, 84, 98, 160, 168, 175–6
saltpeter 120–1, 131–4, 158
saltpeter-men 131–3
Santa Maria Novella, church of 114
Santinus, Martin 194
satires 202–3, 206, 209
Scaliger, Julius Caesar 59
Scatini, Cosmo 119
Scheidung process 29, 48, 55
Schrick, Michael von 146, 148
Scholastics 113
Schroder, Johann 81
science 8, 42, 108
    definition of 67, 92
    new conception of 87, 139
Scientific Revolution 52, 70, 90, 92, 140
"second sight" 8
secrecy 84, 102, 106, 120, 124–5, 140, 160–1, 228
self-transformation 103

Sendijov, Michael (Sendivogius) 178, 182–3, 200
Sennert, Daniel 20, 3, 57–61, 64, 99, 121, 175, 188–90
sense perception 18
Severinus, Petrus 12, 53, 175, 180, 195
Shadwell, Thomas 206–7
siege warfare 130
Sigismund II Vasa, King 183
silver 135–6, 139
Smith, Pamela 142, 152
social mobility 94, 111, 133, 139
society, chemistry in 139–40
Sömmering, Philip 142
Sørensen, Peder *see* Severinus, Petrus
sources of needed materials 43–4
spagyria 25, 50–1, 54–7
Spain 6, 134–5
species 9
Spinoza, Baruch 17
Spranger, Bartholomaus 179
Sprat, Thomas 201
Stahl, Georg Ernst 30, 64, 197
Starkey, George 69, 71, 84, 86
Starkey, Joseph 201
Stradano, Giovanni 126
Stradanus, Johannes 209–11
Strasbourg 146, 149
subcultures of chemistry 92–9, 105, 108
sublimation 69
sulfur 26–9, 52–4, 61, 68–9, 84, 98, 131, 175
Sylvius de la Boë, Franciscus 195–6

Tachenius, Otto 33
Tasso, Torquato 126
Teniers, David the Younger 72–5, 83, 88, 213, 219–21
textile industry 117
Thales of Miletus 30
theory and practice in chemistry 93, 110–11, 168
    barrier between 42–4
    coming together 106
    interdependence of 44
Thirty Years' War 177, 192
Thurneisser, Leonhard 128, 175, 206
Timme, Thomas 134
tin 35